CATASTROPHES IN NATURE AND SOCIETY

Mathematical Modeling of Complex Systems

REM G KHLEBOPROS
Russian Academy of Sciences, Russia

VICTOR A OKHONIN
Toronto University, Canada

ABRAM I FET
Novosibirsk University, Russia

CATASTROPHES IN NATURE AND SOCIETY

Mathematical Modeling of Complex Systems

World Scientific

NEW JERSEY • LONDON • SINGAPORE • BEIJING • SHANGHAI • HONG KONG • TAIPEI • CHENNAI

Published by

World Scientific Publishing Co. Pte. Ltd.

5 Toh Tuck Link, Singapore 596224

USA office: 27 Warren Street, Suite 401-402, Hackensack, NJ 07601

UK office: 57 Shelton Street, Covent Garden, London WC2H 9HE

Library of Congress Cataloging-in-Publication Data

Khlebopros, Rem Grigorievich.

Catastrophes in nature and society : mathematical modeling of complex systems / Rem Khlebopros, Victor Okhonin, Abram Fet.

p. cm.

Includes bibliographical references and index.

ISBN-13 978-981-256-917-2 -- ISBN-10 981-256-917-0

1. Human ecology--Mathematical models. 2. Natural disasters--Mathematical models. I. Okhonin, Victor. II. Fet, A. I. (Abram Ilyich). III. Title.

GF23.M35 K45 2006

304.201'5118--dc22

2006049232

British Library Cataloguing-in-Publication Data

A catalogue record for this book is available from the British Library.

Printed in Singapore by World Scientific Printers (S) Pte Ltd

Dedicated to *Prof. Kenneth R. Kenyon* of Harvard University
who has worked a miracle.

Rem G. Khlebopros

Preface

Complex systems, which defy investigation by the classical methods of mathematical physics, have been among key challenges of science in the recent century. Studies of complex systems primarily address understanding of life and its origin. This mystery prompted Ilya Prigozhine and Manfred Eigen, great scientists and Nobel laureates, to seek the specific properties of living systems that could let an insight into possible ways of how life may have begun. They are presumably self-organization and selection of the most efficient specimen. In fact, the very idea of self-organizing systems with natural selection originally comes from economics rather than from biology and belongs to Adam Smith and Malthus. Malthus's idea of overpopulation encouraged Darvin's discovery of the mechanism that drives the evolution of species.

Self-organization and selection are not specific of living matter but are common to many inorganic systems. They are found, for instance, in convection whirls in fluid and gas which one can observe in a river or in air flows around a hot chimney. However, of special interest are the living systems, from populations of cells in an organism (e.g., cancer cells) or pests in a forest to economic and social systems that appear and disappear in human cultures. In their evolution these systems can develop the states of stable or unstable equilibrium. A system that arrives at an unstable state is prone to a catastrophe — invasion of pests, disease, or crisis — which changes a system dramatically or often drives it to collapse. That is the reason why the book is called *Catastrophes in Nature and Society.*

Classification and prediction of catastrophes is the subject of the catastrophe theory by Rene Thom. Yet, its topological methods cannot provide quantitative predictions so urgent in practical applications. On the other hand, the traditional approaches of mathematical physics implying differential equations are inapplicable to complex systems. The authors of the book approach the subject of complex systems using phase portrait modeling which is closely related to Newton's steepest descent and is thus as classical as differential calculus. Viewed in the context of phase portraits, the processes that seem disconnected and chaotic at first sight show up in their consistent intrinsic logic.

The book written in a popular manner presents the results of original studies by the authors which were partly reported earlier in scientific journals but many come out for the first time in this publication.

The suggested models of mechanisms behind the evolution of nature and society are of great theoretical and practical value. Their understanding and proper use can reduce the risks that threaten our civilization and eventually bring to harmonic coexistence of economy and environment instead of their today's antagonistic confrontation. One can expect that the book will make for better awareness of readers willing to participate in solving the vital environmental, economic, and social problems of our time.

Professor S. Gabuda

Acknowledgements

We wish to express our gratitude to people who contributed to the preparation of the manuscript.

V.A. Slepkov wrote Chapter 2 on the basis of studies run jointly with Rem Khlebopros. We also thank him for discussions we profited from when writing Chapter 9.

V.G. Suhovol'sky was the first to apply the Zipf-Pareto law to check votes count in elections and wrote the respective section in Chapter 15. We appreciate his criticism of the remaining part of Chapter 15 which is of our full responsibility.

V.V. Mezhevikin was very helpful with consulting on biochemical issues essential for better exposition of Chapter 6 on manmade closed ecosystems.

V.B. Kashkin furnished global total ozone data and developed the experimental method as a basis of Chapter 5.

Comments by D. Proy prompted us to make a number of useful improvements in Chapters 1–3 and 6–8.

The manuscript benefited much from constructive criticism by I.I. Gitel'zon.

Special thanks go to T.I. Perepelova for her aid in editing and making the English text.

Contents

Chapter 1

Environmental Microcatastrophes

This opening chapter presents an approach which is often used to investigate the behavior of complex systems when the number of equations that allow for interactions of their many parameters is too large and the direct methods are impracticable. A good way to illustrate the approach is to apply it to a forest ecosystem in terms of its resistance against pest insects and environmental microcatastrophes associated with insect outbreaks. They are called "microcatastrophes" as it is a single ecosystem that runs a risk of destruction.

Population outbreaks, in which populations suddenly increase many times, have occurred in nature long before the appearance of man, for instance, the well-known disasters such as plagues of mice, lemmings or locusts. Man interferes with the natural dynamics of animal populations, being able to both provoke and stop outbreaks. In some cases man-induced outbreaks can result from introducing a new species to a country that naturally lacks conditions to control its numbers (say, natural enemies). This was the case of rabbit brought to Australia by European settlers, or Colorado beetle, a potato pest, accidentally brought to Europe and Asia. On the other hand, there exist chemical and biological methods of mitigating risks from insects and other pests.

Forest pests offer a classical model of natural population outbreaks, such as, for instance, outbreaks of black fir beetle (*Monohamus urussovi* Fish.) that inhabits the Sayan Mountains in Siberia. At the beginning of its outbreak, the population of insects in a forest is very small but then it grows suddenly, increases tenfold for a few years, and sweeps over the forest to destroy a great part of trees; eventually, the outbreak fades and the population returns to its previous moderate numbers.

We study the important problems of population dynamics, such as stabilization of populations, cyclic processes in population density, and population outbreaks, using a very simple model. Although simple, it can describe and predict outbreaks of insect populations just by means of straightforward plotting without any resort to calculations.

Phase portraits

Prediction of the population density in a coming year from its density in a current year is a key point in studies of population dynamics. If the number of insects of some species living in some area is K in a current year and M in the following year, the ratio M/K measures the growth (if $M/K > 1$) or decline (if $M/K < 1$) of the population. The ratio M/K is called the reproduction rate of the species in a given habitat. The population number is proportional to the surface area and M/K is constant throughout the habitat if individuals are evenly distributed and the environment is more or less uniform. In such cases population dynamics can be studied in selected sample areas of, say, one square mile. In fact, M/K may differ through different parts of the forest and, what is most important, through different years. The current number of insects obviously influences reproduction as it controls the availability of food and habitat resources. For example, a too large population (overpopulation) slows down its reproduction because less progeny can survive in the conditions of severe competition.

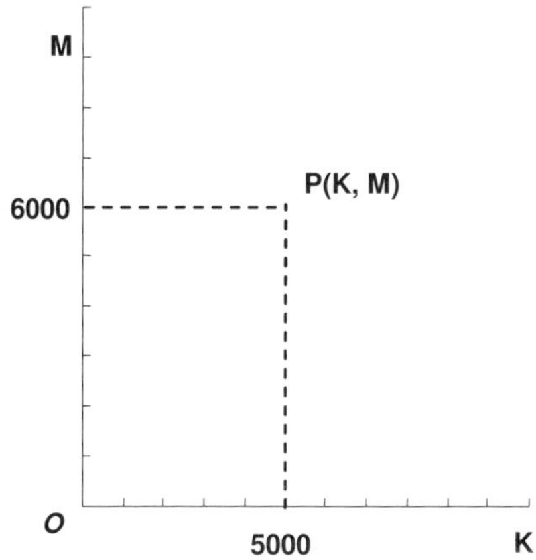

Fig. 1. A single standard observation on the phase plane (K, M), where K is the number of insects in a year, M is the number of insects a year after.

Thus there is no simple rule to predict the population density in some year from its density in the previous year. The problem can be solved using the method of phase portraits. Phase portraits are plotted in the Cartesian coordinates as points in the phase plane with the coordinates made by pairs of numbers (K, M) which are called standard observations and obtained by counting insects of a given species living within a given area in a current year (K) and insects in the same area in the following year (M). The K and M axes are assumed to have the same scale (Fig. 1), for example, 1000 individuals. A standard observation is represented in Fig. 1 by the point P with the coordinates (K, M), with $K = 5000$ and $M = 6000$, or 5000 individuals in a current year and 6000 insects a year later.

The totality of standard observations for several years (in the same or different areas) plotted in the phase plane makes up a "cloud" of points (Fig. 2), which is called the *phase portrait* of the respective population.

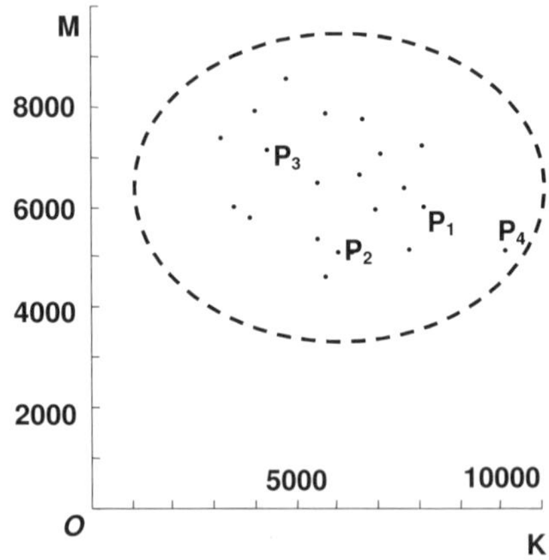

Fig. 2. Phase portrait of a population comprising standard observations for several pairs of successive years.

Each point in the cloud represents two counts (*K* and *M*) of population density in two successive years (Fig. 2). For instance, the point P_1 corresponds to 8000 insects counted in 1995 (*K* axis) and 6000 insects in 1996 (*M* axis) and P_2 shows the population change from 6000 insects in 1996 to 5000 in 1997; P_3 and P_4 are for 4000 insects in 2000 and 7000 in 2001 and 10000 insects in 2002 and 5000 in 2003, respectively, etc.

Even though different points of the cloud can correspond to the same area (and different years), in the general case they are related to different areas and image the results of numerous long-term observations each taken at two successive years. A question arises whether the data obtained this way represent any dependence between the population numbers *K* in the current year and its numbers *M* in the next year on the same plot. Generally, there is no such dependence as any *K* can have more than one *M* in the cloud because besides the initial number of insects their reproduction depends on the environment, weather, or living quality of the selected habitats.

However, in many important cases environment factors are less essential than the initial population density *K*, and the cloud of Fig. 2 then becomes elongate (Fig. 3*a*). The cloud stretching and narrowing (provided that observations are complete) mean that the population in the following year ranges within a narrow interval (M_1, M_2) for any specific *K*.

A cloud which is narrow enough can be approximated by a curve (Fig. 3*b*). This is an approximate phase portrait giving a tentative idea of

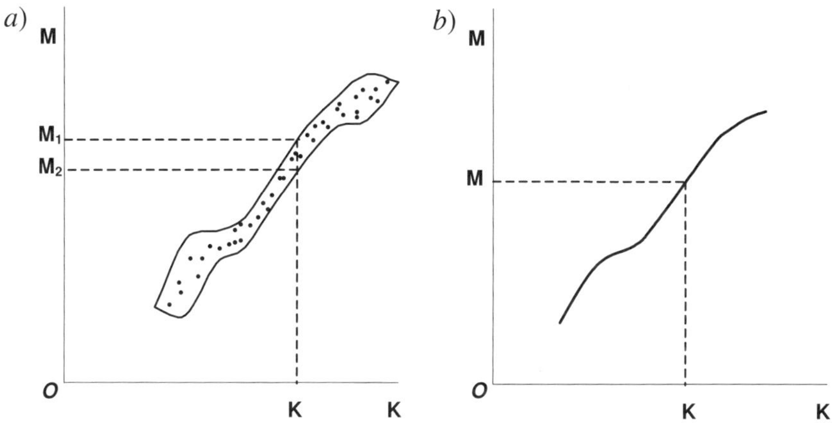

Fig. 3. Narrow phase portrait of a population (*a*), and its approximation by a curve (*b*).

reproduction without allowance for environment and climate controls. Thus any *K* corresponds to a single *M* (Fig. 3*b*) and their relationship is a function where *K* is the argument and *M* is the function.

In simple cases, functions are defined by equations such that a function is easily obtained from its argument ($M = K^2$, $M = (K+1)^3$, or $M = \log K^2$, etc.). These equations are usually mathematically proved or obtained from physical theories for simple natural phenomena. Yet, the processes challenged in environment science (or in other sciences such as biology, economics or sociology) are often too confusing to be described by mathematical equations but can be approximated by empirical relationships [*Kolmogoroff*, 1937, etc.] like the curve in Fig. 3*b*.

Substituting a curve (Fig. 3*b*) for a narrow cloud (Fig. 3*a*) is a very common procedure applied to any empirical function (obtained from experiment). The cloud of standard observations can be in fact much narrower than that shown in Fig. 3*a*, and the approximation 3*b* is in this case more exact. We assume this approximation to be valid in population dynamics problems, as well as other problems of environment, sociology, and economics discussed below. Of course, this assumption requires a support from experience, and it does agree with practice quite often.

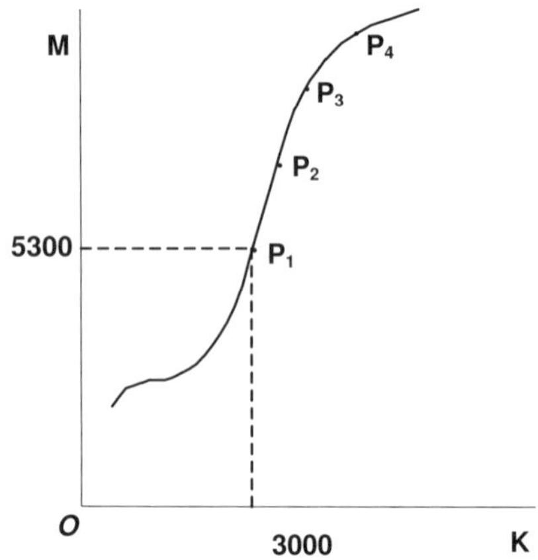

Fig. 4. Standard observations P_1, P_2, ..., plotted to make an empirical reproduction curve considered as the phase portrait of a population.

The phase portraits of processes (e.g., see the series of points P_1, P_2, ... , etc. in Fig. 4, standard observations over many years making up an empirical reproduction curve) are plotted assuming that the processes can be described by clouds almost coinciding with a curve. Thus the phase portraits considered below are actually the phase curves, or the plots of some functions.

The most general trends are evident already from the very shape of phase curves (Fig. 5). For instance, curve 1 everywhere rises (greater M correspond to greater K) and is an increasing function $M(K)$, whereas curve 2 everywhere falls (greater K corresponds to smaller M) being a decreasing function. The function of curve 3 first decreases and then increases having reached its minimum. Similarly, it is easy to plot a function with a maximum.

Curve 1 is convex. This means that any chord connecting two points cuts off an arc above the chord. A convex curve looks like an upside-down cup. Curve 3 is concave, i.e., its any chord truncates it to make an arc below. A concave curve looks like a right-up cup. Curve 2 is first concave and then convex, and its concave and convex segments are separated by a point of inflection.

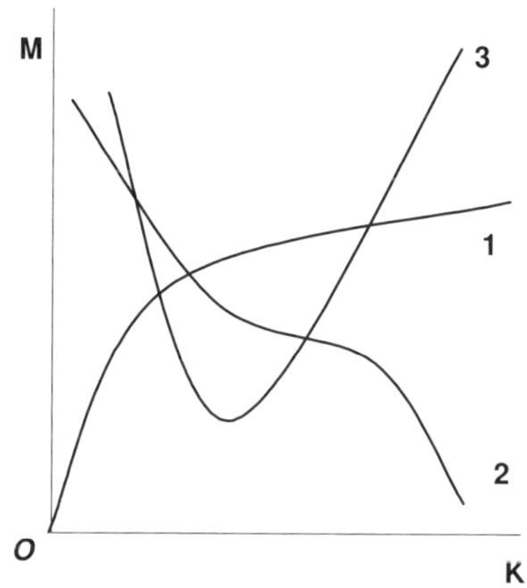

Fig. 5. Most general types of reproduction curves: increasing curve (1), decreasing curve (2), concave curve with minimum point (3).

Reflection from bisector

Phase portraits obtained from long-term observations can be used to predict changes many years in advance. This is best achieved by a straightforward geometric device applicable to many problems other than population dynamics.

Many populations have their phase portraits in the form of a convex curve that begins from the origin of coordinates *O* and crosses the bisector of the quadrantal angle at a single point (point 1 in Fig. 6).

Consider this common case in more detail. The origin of coordinates *O* represents the simplest and a quite natural standard observation when there are no insects at all in a current year ($K = 0$) and, naturally, no insects in the following year ($M = 0$). Of course, repeated observations drive to the same result, as predicted by the phase portrait: zero population in any current year and zero population in the following year.

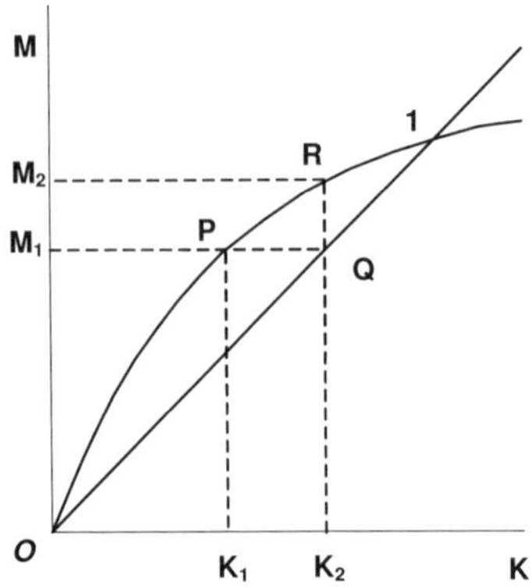

Fig. 6. Method of reflection from bisector applied to a convex reproduction curve.

This means that the population is extinct. A point on the phase curve where the population is constant ($M = K$) is its equilibrium point and the corresponding state of the population is its equilibrium state. Therefore, the bisector which is the totality of all points $M = K$ is of special importance to our further studies. The plot in Fig. 6 has a single equilibrium point besides O where the phase curve crosses the bisector (point 1). Let it have the coordinates K_s, M_s where "s" stands for "stable (equilibrium)"; then, $M_s = K_s$, i.e., the number of individuals (K_s) in a current year is exactly the same as in the following year (M_s), etc. Such exact equality is of course rare in practice but K can approach K_s. We show later what comes from that, and now consider the population a year after the first year of observation. It can be predicted from M_1 corresponding to K_1 (K_1 is taken positive but less than K_s. See the phase curve of Fig. 6). The same population another year later is predicted by repeated use of the same phase curve: It is the M-coordinate M_2 of the point R with K_2 equal to M_1, as M_1 is then the initial population, and M_2 is the population a year after it was M_1. To simplify the transition from P to R, see that both OK_2 and OM_1 are the sides of a square with the vertex Q on the bisector (as $OK_2 = OM_1$ and the bisector is the locus of points equidistant from both axes). Thus, the point Q can be easily found by drawing a horizontal line through P to cross the bisector. The point R has the same K_2 as in Q, and R is thus obtained by drawing a vertical line through Q to cross the bisector.

Therefore, the population M_2 a year after it was M_1 can be inferred geometrically: to pass from the point P of the phase curve to the point R, simply draw a horizontal line through P to cross the bisector in the point Q and then a vertical line through Q to cross the phase curve in R.

Then only M_1 and M_2 are to be used, without plotting K_1, K_2 and the segments K_1P and K_2Q (Fig. 6). To predict the population M_2, it is enough to plot the rectangle M_1QRM_2 knowing M_1. The segments PQ and QR are at equal angles to the bisector (45°), like the light reflected by a mirror. That is why the device is called reflection from bisector.

Simple population processes

We begin with simplest processes described by increasing functions. Fig. 7 images a common and very important case of $M(K)$ relationship represented by a convex phase curve starting from the origin point O and crossing the bisector at a single point (point 1). The coordinate origin ($K = M = 0$) is an equilibrium point, but not an interesting one since the population does not exist there at all. The point acquires special importance when P approaches O and the population is extinct but it is evidently not the case for the considered type of the curve since population density at the interval above the bisector R is always to the *right* of P, and $K_2 > K_1$ (Fig. 6). Moreover, due to the convexity of the curve its chord OP goes down when P moves to the right (see Fig. 7) and, therefore, the slope of OP relative to the K-axis also decreases (the slope is measured by the ratio of the ordinate (M-coordinate) of P to its abscissa (K-coordinate) and equals the tangent of the angle of OP to the K-axis). Thus, the ratio M/K diminishes when K grows. The ratio, which is the reproduction rate of a population, measures how many times the next generation is larger (if $M/K > 1$) or smaller (if $M/K < 1$) than the previous one. On the contrary, the reproduction rate grows when K diminishes, and reaches the highest value at $K = 0$ associated with biologically vanishing population density. Mathematically, the highest value is achieved through passage to the limit: as K tends to zero, the slope ratio of the chord OP, equal to M/K, tends to some limit called the derivative of the function $M(K)$ at $K = 0$ and denoted as $M'(0)$. In terms of geometry the limit is the slope of the tangent to the curve at the point O.

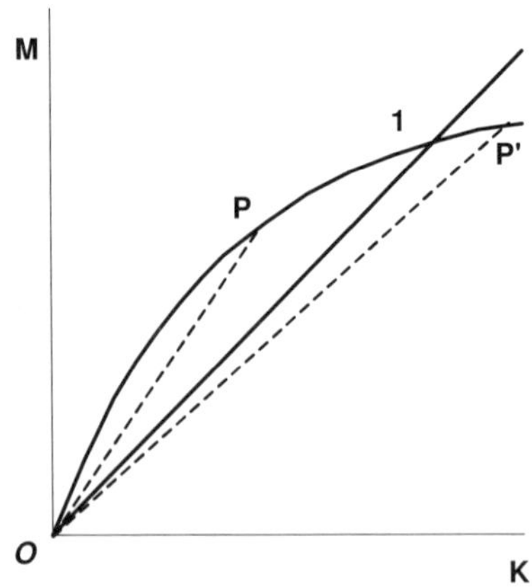

Fig. 7. $M(K)$ function represented by a convex phase curve which includes the origin point O and crosses the bisector at a single point (point 1).

The biological meaning of the result is that the reproduction rate of a population decreases after it has occupied the best habitats for feeding and laying eggs. Slower reproduction can also result from other factors such as contamination of environment with the population's life wastes, or infectious diseases. The competition for habitats might serve to avoid the negative effects. The ratio M/K is above unity till point 1 (at $K < K_s$ where K_s is the K-coordinate of 1) (Fig. 7) which means that $M > K$ and the population grows year by year. However, at $K > K_s$ the slope of OP becomes below unity, i.e., $M < K$ and the population reduces, which can be interpreted as the overpopulation effect [*Baltensweiler*, 1964; 1970; *Baltensweiler et al.*, 1977].

Equilibrium states corresponding to the points $M(K) = K$ are of special interest. Once fallen at this point, the population obviously stays there forever. Of course, the result has only theoretical significance, since any casual disturbance can push the system out of the equilibrium state. That is why we concern mainly the states of stable equilibrium which a population tends always to return to after minor deviations. In our case the convex phase curve can cross the bisector at only one point. Let it be point 1, its K-coordinate be K_s, and its M-coordinate be $M_s = K_s$.

We show below that point 1 is a point of stable equilibrium of a population.

Let the point P_0 (K_0, M_0) in Fig. 8 mark the state of a population having its initial density K_0 (K-coordinate of P_0) and the density M_0 in the next year (M-coordinate of P_0). The projections on the axes are omitted to make the picture better readable.

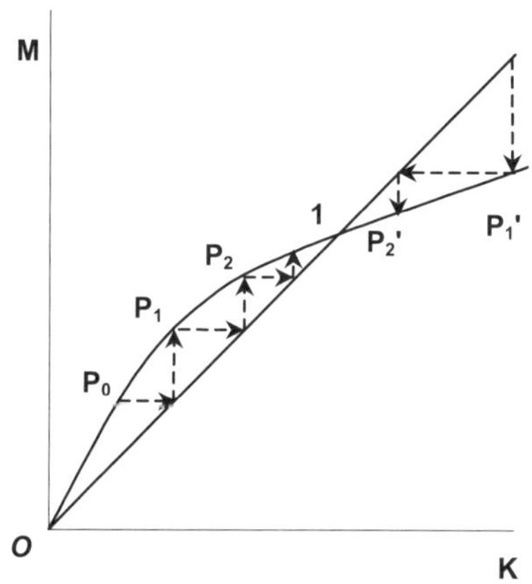

Fig. 8. Population dynamics for a convex phase curve.

At $K_0 < K_s$ (P_0 is on the left of 1), the reflection from the bisector moves P_0 into P_1 imaging the population state in the next year with the population density equal to the K-coordinate K_1 of the point P_1 (Fig. 8). Repeating this procedure gives P_2 with the K-coordinate K_2 equal to the population density in the following year after P_1, etc. The points P_0, P_1, P_2,... arbitrarily approach point 1, so any population that initially was on the left of 1 evolves to approach the point of stable equilibrium (point 1) arbitrarily close (Fig. 8). On the other hand, if the initial state P_0' was on the right of 1, it grades into P_1', P_2', P_3',..., approaching 1 from the right. The effect of some casual factors can move the point P off or make it leap over state 1, but the population tends to recover its stable state from either direction. This is the phenomenon called a state of stable equilibrium of a population: whichever be its initial distance from the point of stable equilibrium, the population eventually stabilizes near it (at the type of the curve $M(K)$ we discuss). Therefore, the population density is always near K_s unless some extraordinary disturbance occurs.

In another case, as simple as the previous one, of a concave $M(K)$ crossing the bisector at a single point (point 1, Fig. 9), any population state P_0 between 0 and 1 successively grades into the states P_1, P_2,..., approaching 0, and the population vanishes. If the initial state of the population P_0' is to the right of 1, its following states P_1', P_2',..., correspond to the tendency of an unlimited growth. This situation is obviously impossible in nature, and the curve of Fig. 9 thus does not provide a realistic image of the reproduction dynamics at high K; it can no longer be concave (see below).

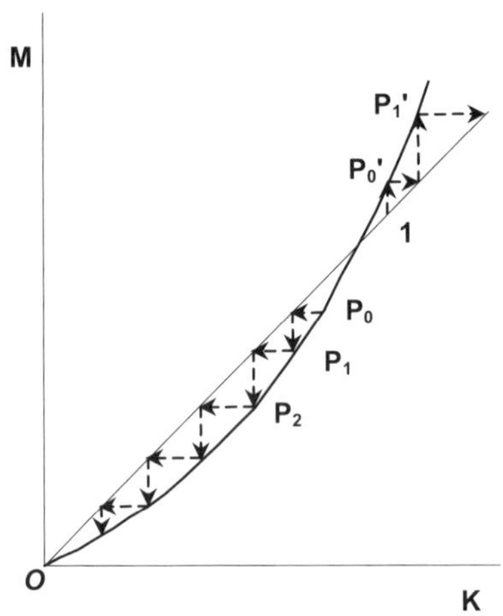

Fig. 9. Population dynamics for a concave phase curve.

Point 1 is again an equilibrium point, but in this case it is a point of unstable equilibrium when any minor deviation drives the population either to extinction (deviation to the left of the equilibrium) or to unlimited growth (deviation to the right of the equilibrium). This state obviously cannot persist and thus eludes any observation in nature. These points are important rather because they are the "divides" between different domains of population dynamics [*Isaev and Khlebopros*, 1973; *Nature, Our Soviet correspondent*, 1973].

The curve of Fig. 8 and, hence, populations of stable density, are commonly found in nature at invariable conditions. The curve of Fig. 9 is, on the contrary, unrealistic in general, and only its concave part is essential showing the decay of a population towards the left end.

More complex population processes

So far we have focused upon increasing functions $M(K)$ and thus could predict only the dynamics associated with population growth or decline till its equilibrium state (including zero point corresponding to extinction). Yet, some species demonstrate a different type of population dynamics, likewise existing in nature, and their phase functions are not always increasing. The assumption that a population decreases at high K means overpopulation and can be checked in the field.

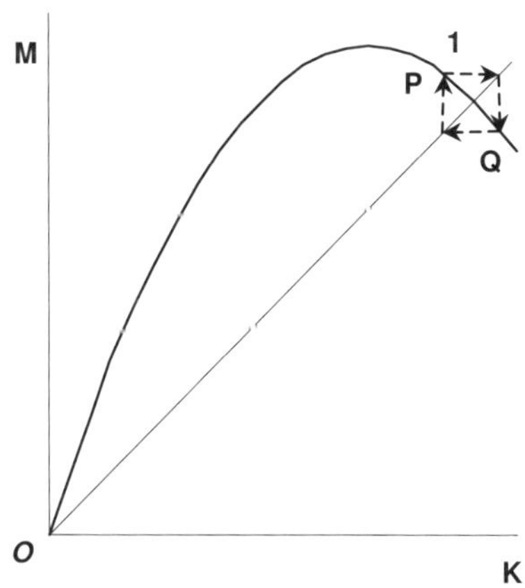

Fig. 10. Idealized closed cycle for a convex phase curve.

Figure 10 shows the curve $M(K)$ with an interval of rapid decrease after equilibrium point 1. A population that happens to be in the state P (or Q) (the vertices of the square halved up by the bisector), is in the state Q (or P) the following year and so on, and its density oscillates periodically.

This "exact" cycle is of course impossible in reality, because the initial state is never exactly P or Q. A natural process of this kind can be interpreted as cyclic only to some approximation. These nearly cyclic processes actually get disturbed with time. Figure 11 images the density dynamics of a population pushed off its cyclicity and evolving in a spiral way. The K-coordinate of K can vary in a rather broad range about the equilibrium (K-coordinate of 1). These processes are called quasi-chaotic, and this equilibrium point is called unstable since the population density can significantly deviate from the equilibrium infinitely many times.

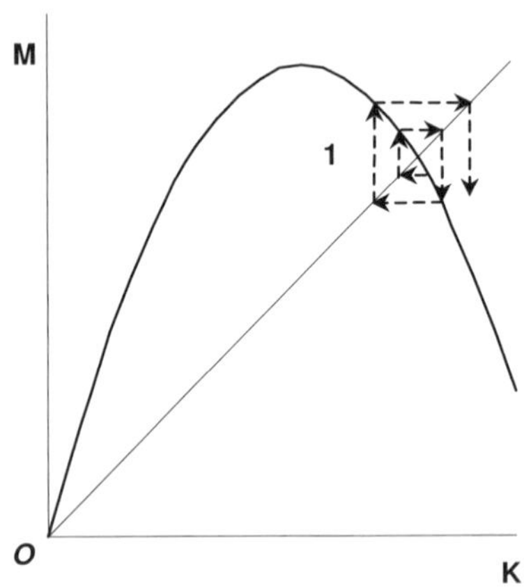

Fig. 11. Quasi-chaotic population dynamics for a convex phase curve.

Many natural processes behave so that their phase curves cross the bisector at three points 1, 2, 3 (excluding the origin O), which is the case of special importance. In this case population density has three equilibrium points, except zero where it becomes extinct. Figure 12 shows the most common case of an increasing phase function $M(K)$.

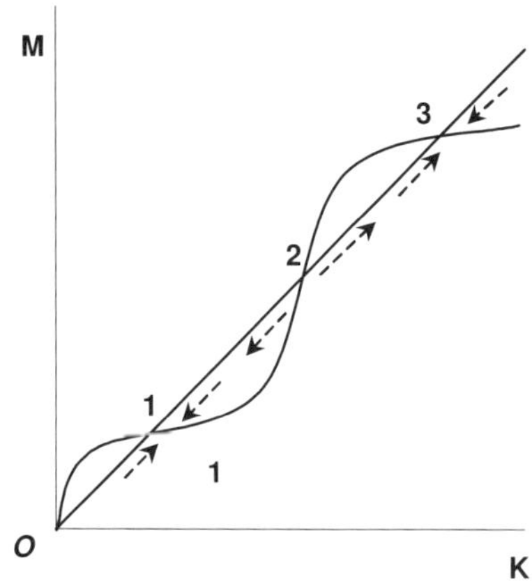

Fig. 12. Phase curve that crosses the bisector at three points of stable (points 1 and 3) and unstable (point 2) equilibrium.

We turn to the biological meaning of the curve of Fig. 12 after having formally applied reflection from bisector.

The initial states between points 0 and 1 move to the right approaching point 1, and the initial states between points 1 and 2 move to the left, likewise approaching point 1. Therefore, point 1 is again the point of stable equilibrium, with the only difference that a too strong deviation towards population growth makes the population move to the right rather than turn back to point 1, since the states between 2 and 3, that likewise are above the bisector, move rightward like the points between 0 and 1. So, point 2 marks unstable equilibrium as in the case of Fig. 9. Point 3, on the contrary, corresponds to stable equilibrium, like point 1, and is associated with a higher population density relative to point 1. Remember that stability of a population is meant as "local", and the system never recovers after too large deviations from the equilibrium.

We can easily apply the same constructions as in Figures 8 and 9 to the more complex curve of Fig. 12 at intervals 0-1, 1-2, 2-3, and the interval to the right of 3.

The interval 0-1 is favorable for a species survival as any population within it grows till point 1 of stable equilibrium where it stays (slightly oscillating around it due to casual effects). Interval 1-2 is unfavorable and is associated with population density decrease till stable point 1.

Interval 2-3 is again favorable and brings the population to stable equilibrium 3. If the curve does not cross the bisector at other points after 3, all states to the right of 3 approach it while the population decreases. Further intersections, if any, are alternately the points of stable and unstable equilibria. As far as we know, natural insect populations never show more than three equilibria (except zero).

A question naturally arises: which biological conditions control the specific shape of the phase curve of a population? These conditions are known, for example, for the forest pest of black fir beetle. The curves of the type of Fig. 8 describe its populations in young forests in plainland or in forests at the steppe border in highlands. Black fir beetle attacks weak and ill trees in those forests by getting into their inner bark where it lays eggs. The pest does not attack strong trees which protect themselves by coating the eggs with gum. Equilibrium point 1 in Fig. 8 corresponds to the pest population of stable density that live in a forest with a constant percentage of weak trees to provide food and breeding place.

The conditions of higher mountains, where climate is colder and wetter, are more favorable, and the initial population density K is associated with a larger density M in the following year, so that the phase curve of Fig. 8 rises up to take the shape of Fig. 12. Interval 2-3 corresponds to another feeding pattern specific to the case when the population density is so high that pests have to attack sound trees for the shortage of weak ones, and sound trees can no longer resist. Unlike the case of infinitely enduring ecological equilibrium between the forest and the pests in plainland and low highlands, the new feeding pattern threatens the very existence of the forest. Yet, insects can use interval 2-3 only after their density reaches point 2. Normally this never happens and the population stays in some vicinity of stable equilibrium 1 since it cannot grow within 1-2. Abnormal conditions arise when a highly dense population comes from outside, which is quite possible for flying insects. Then the density of pests grows at the account of the invaders rather than the natives who are unable to overcome the unfavorable interval between 1 and 2. As soon as the population exceeds point 2, interval 2-3 “sets into work” and the population density rises up to new stable equilibrium 3 to produce an outbreak. Population density at state 3 may be ten times that at normal state 1 in black fir beetle and dozens thousand times in other insects, which makes interval 2-3 too long to show in a picture.

The conditions are especially favorable in mountain top regions (in low highlands), where a mature forest has a different phase curve than in Fig. 12, namely it rises relative to the latter (Fig. 13). Biologically the upward shift means that the same population density *K* in a mature forest produces a higher density *M* the following year than in the worse conditions of Fig. 12. The risen phase curve crosses the bisector again at a single point (point 3) while points 1 and 2 disappear.

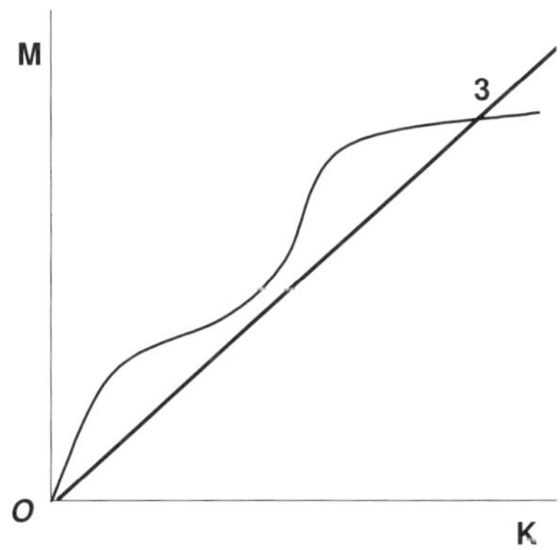

Fig. 13. Phase curve associated with the conditions most favorable for pests or with a weakened forest.

A pest population with this phase curve reproduces till point 3 and an outbreak becomes inevitable. Once the population becomes dense enough, insects attack mature forest and that feeding pattern persists until the whole forest dies out. Driven by the food lack, the dense population migrates downhill and triggers the formerly inactive interval 2-3 of the curve of Fig. 12. Then the pests develop an outbreak which propagates downhill like a wave destroying the forest. Finally the outbreak reaches its lower part corresponding to the phase curve in Fig. 8. The population cannot reach a too large density there since it can only diminish within the region to the right of 1; the excess insects die out and the outbreak stops [*Isaev et al.*, 1984].

Pests usually spare young trees, and the forest thus can recover after outbreaks, starting with the age relevant for the insects. Then the tops and slopes of mountains become again covered with mature forest, roughly uniform in age, where another outbreak can occur in a while. The periodic outbreaks devastate forests in the same way as forest fires, with the only difference that forests can recover after outbreaks of pests and fires can often be fatal. In such cases, the mountains get covered with grass, or with a forest of a different type. That is why overmature forest with trees of different ages is never found in upper mountains where air or satellite images most often show young forests.

Interference of man can deform the phase curve, most often to the worse. In nature the curve of Fig. 13 rises under conditions that are especially favorable for insects when they can feed on strong trees. Yet, the same thing happens in a forest weakened by felling, or by industrial air pollution. This is the way man can provoke insect outbreaks.

On the other hand, man can mitigate the risk of outbreaks, or stop them, by means of chemical or biological extermination of pests; then the phase curve goes down, as in Fig. 14, and returns to the form of that in Fig. 8.

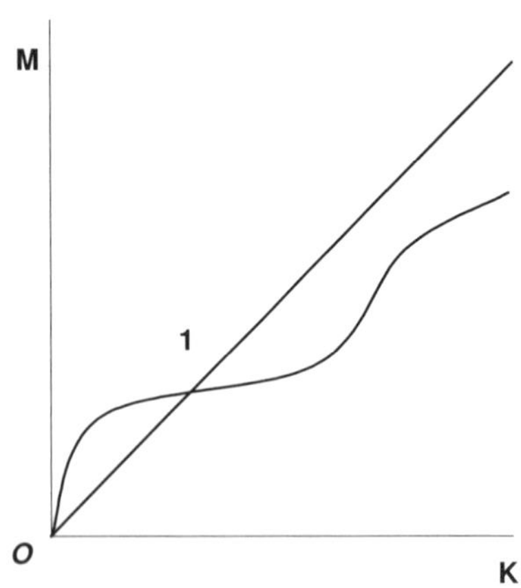

Fig. 14. Phase curve associated with manmade extermination of pests.

Up to this point, we have dealt with population outbreaks of pests and their interaction with the host forest, but the model has more general implications. A few examples below show outbreaks relevant to people and their activity.

Locust outbreaks are the most familiar. In terms of taxonomy, the locust is a species allied to grasshopper living in arid steppes, particularly in the steppes of Eastern Asia. In normal conditions they are not social insects, and live like ordinary grasshoppers. However, if their density becomes excessive for any reason (not quite understood as yet), they grow wings, normally reduced, and acquire the ability to fly. Then, gathering in immense swarms, they become what is called locusts, and fly to other countries, devastating vegetation, and migrate on and on as far as they eat up all food, exactly as the forest pests do.

Another example is the vanished forests of Ancient Greece. Throughout historic time, this mountainous country has been woodless, with bare rocky ranges. However, the mountains were covered with forest in prehistoric times, as is remembered in the Greek oral tradition and confirmed by paleogeographic data. The scarce population of that time could hardly disforest the land. There is a hypothesis that the deforestation may have been caused by domestic goats which ran wild and went through an outbreak. The goats ate up the young tree suckers, the forest never recovered, and rains washed the soil away from the slopes. Thus, the familiar classical Greek landscape results from a man-induced environmental disaster. The same impact, for instance, related to exhaustive pasturing, can provoke desertification. Perhaps, Sahara might have appeared in this way.

We can also cite an example of rabbits that were brought by immigrants to Australia where they could not find enough natural enemies since the population of local predatory marsupials had been thinned out at that time. Consequently, the uncontrolled rabbit population grew in an outbreak that endangered the entire agricultural production of Australia. Eventually people had to build fences throughout the continent to keep the rabbits away.

The species of *homo sapiens* may have likewise undergone an outbreak when people learned to hunt mammoths and thus passed to a new feeding pattern. When mammoths became extinct, the human population may have considerably decreased. Another outbreak of human reproduction apparently occurred in the twentieth century.

A case of erroneous ecological measures

The reproduction of salmon provided an instructive example of misprediction and related erroneous measures. Adult salmons inhabit open sea but come to spawn to the rivers of the Russian Far East and America. Each seven-year-old fish returns to its birthplace where many fish gather for spawning. The Russian zone of catch was established as a 12 miles wide border where only Russian fishermen were allowed, but Japanese fishermen fished extensively along the border margins. To stop that practice, ichthyologists suggested the government to extend the boundary up to 24 miles, assuming that this would automatically increase the yield. However, the catch actually decreased and increased only later, after several seven-year cycles. To explain what actually happened, we interpret the reasoning of the ichthyologists (who used a different approach) in terms of our model (Fig. 15), where population numbers K and M correspond to two successive septennials and thus are spaced at seven years. (The model is the same whether we count fries or fish returning to spawn).

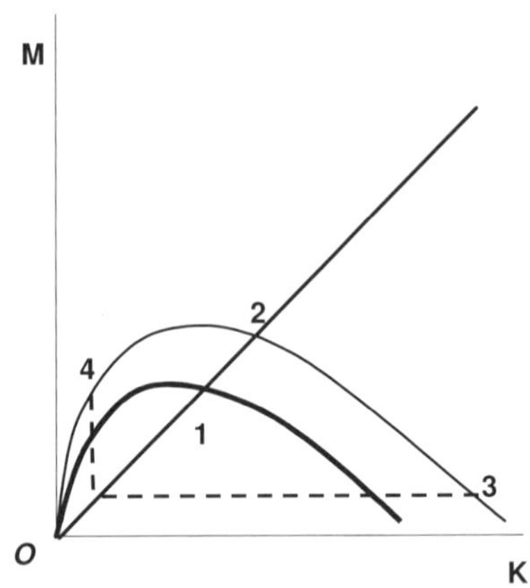

Fig. 15. Phase portraits of the salmon population before (heavy line) and after (thin line above) the broadening of the forbidden catch zone.

Salmon is a species which decreases its numbers abruptly at overpopulation, i.e., its phase curve bends sharply downward. The phase curve shown by heavy line in Fig. 15 corresponds to the catch before the broadening of the forbidden zone and the curve shown by thin line to that after the change. In the former case the population was at the state of stable equilibrium 1; in the latter case it was presumed to pass to a new equilibrium state 2, with a greater population.

In fact, the catch fell strongly only for the fish coming to spawn because the change of the forbidden zone did not apply to the catch in open sea or in rivers. As a result, much more individuals could successfully reach the spawning place. Describing the new situation in terms of phase portraits requires primarily the knowledge of what would be the initial population at the point where we start with the new phase curve. We may assume that the fish density in the spawning places would be excessive, leading to reverse dependence between the progeny and the numbers of spawning fish. Therefore, fish, in fact, hindered one another in spawning so much that the damage could not be kept up by the greater numbers of spawning individuals.

The first reproduction cycle after broadening the forbidden zone is represented by point 3 of the new phase curve, where K is much greater than the former stable number of fish coming to spawn (the K-coordinate of point 1) However, after seven years we obtain a number M smaller than the previous stable value. One might predict that after several further septennial cycles (first such cycle corresponds to point 4 of the new phase curve), the population should stabilize at point 2, and thus become greater than at the former point 1. That is what really occurred.

Classification of outbreaks

Typical phase portraits allow classification of all possible kinds of outbreaks understood as any sudden change in population density. We limit ourselves to the cases when the curve $M(K)$, starting at the origin of coordinates, crosses the bisector at three other points: point 1 corresponds to the lowest nonzero population density, point 2 to an intermediate population density, and point 3 to the highest population density. Each of the three can be, in principle, stable or unstable, but only certain combinations of stability types are possible. Our assumption agrees with the fact that no more than three equilibrium states are observed in nature, at least in insects.

Intermediate equilibrium point 2, even if it is stable against small deviations, rarely holds on for a long time, and thus this type of population is rare for long-living species. Indeed, suppose that such a point is stable, as shown in Fig. 16.

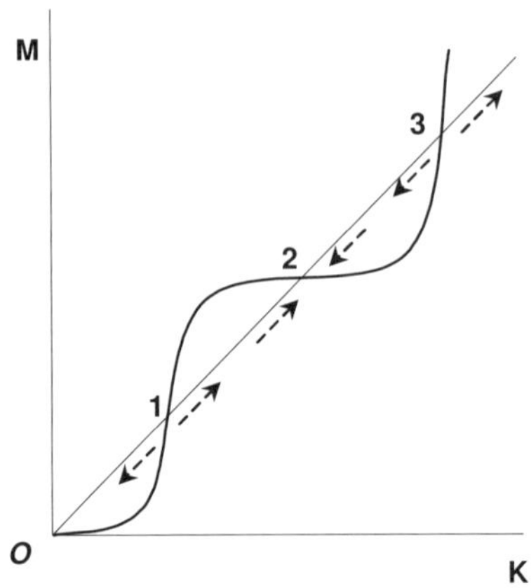

Fig. 16. Phase curve that crosses the bisector at three points: lowest nonzero population density (point 1), intermediate population density (point 2), highest population density (point 3).

The zero point is also necessarily stable in this case (Fig. 16). Of course, a population that fits this phase curve arrives at point 2 and can never pass the unfavorable zone 1–2 downwards. However, some events can upset the normal behavior of the population and suddenly increase or decrease its density, as in the case of flying insects brought to some locality, to move the population to the right of point 3 into the interval which was out of reach before. In the same way, some disaster can push the population of Fig. 16 to interval 0–1, make it move to the point *O* and eventually become extinct. Therefore, species with the dynamics corresponding to Fig. 16 are relatively rare, and can be neglected in rough classifications like the one we suggest. (Compare this case with the behavior corresponding to the phase curve of Fig. 8 when a population never fails to return to stable equilibrium 1, even after great deviations.)

The population of species close to extinction usually decreases too much in density, as is probably the case of whales because of abusive hunting and increasing pollution of the oceanic water. There are, however, species able to prevent population thinning by means of the so-called Allee effect. They form geographically dispersed groups with a relatively dense population in each, which keeps them within interval 1-2 of the phase curve, i.e., above critical point 1. The Allee effect is worth mentioning as it extends Lorenz's general law which implies that the instinct of intraspecific aggression forces species to settle uniformly over their habitats. Note that the curve of Fig. 16 must bend downwards to the right of point 3 because of the limited resources and the related constraints on population growth.

No fourth equilibrium point having been so far found in insects [*Isaev and Khlebopros*, 1977; *Berryman et al.*, 1987; *Bazlkin* et al., 1997; *Isaev et al.*, 1984; 2001], we may restrict ourselves to the cases of unstable point 2. Then there are several alternatives[a].

Points 1 and 3 correspond to stable equilibrium [*May*, 1971; 1973; *Isaev and Khlebopros*, 1973; *Nature, Our Soviet correspondent*, 1973] (Fig. 17):

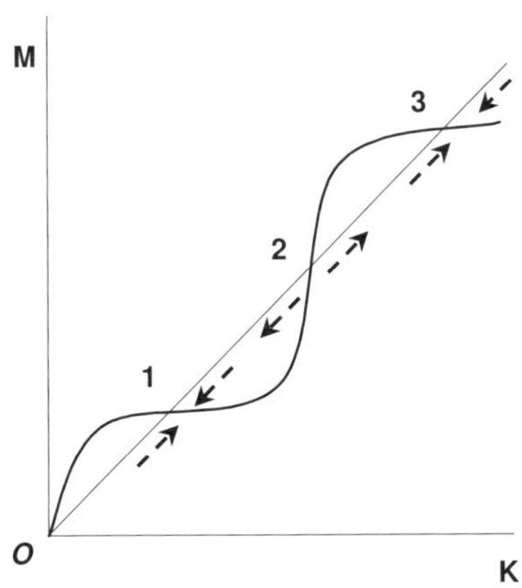

Fig. 17. Case of stable points 1 and 3.

[a] They can be investigated using reflection from bisector in the same way as above.

Point 1 is stable, and point 3 is unstable (corresponding to unstable equilibrium) (Fig. 18). Note that explanation to Fig. 11 shows why point 3 is unstable.

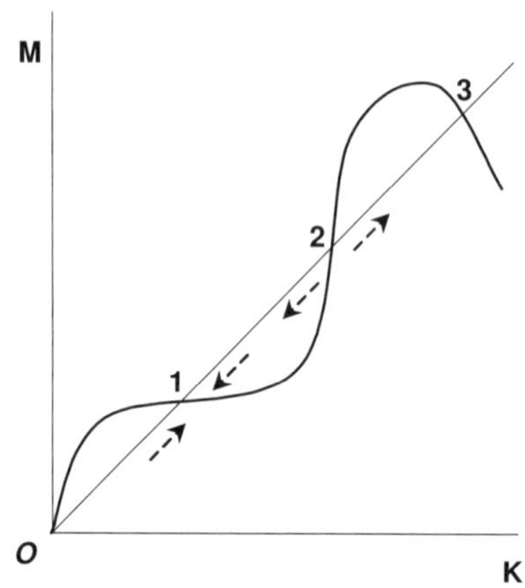

Fig. 18. Case of stable point 1 and unstable point 3.

Point 1 is unstable and point 3 is stable (Fig. 19):

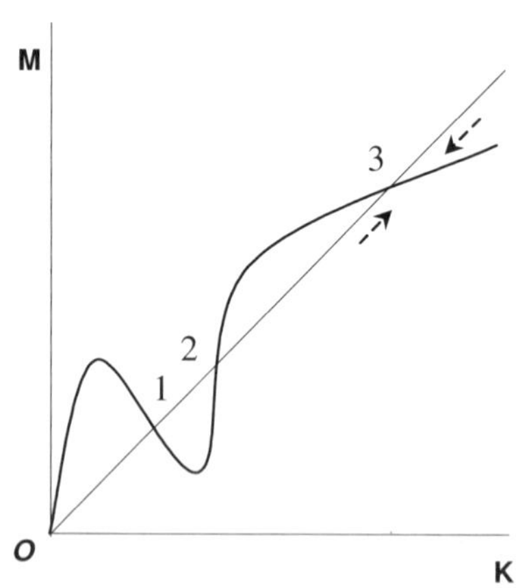

Fig. 19. Case of unstable point 1 and stable point 3.

Points 1 and 3 are both unstable (Fig. 20):

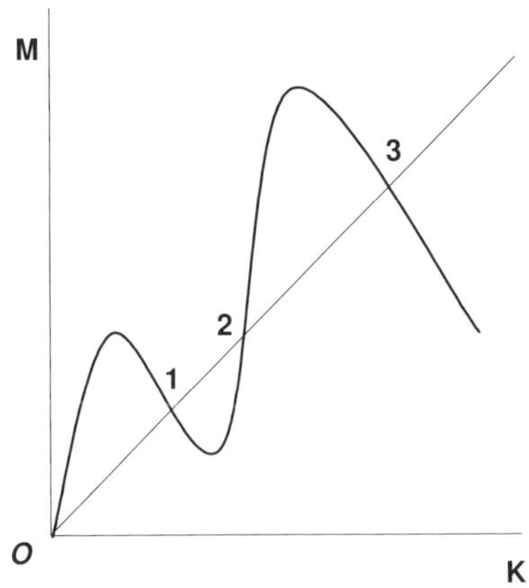

Fig. 20. Case of unstable points 1 and 3.

Figures 18, 19 and 20 image the cases when a population moving from unstable points 1 or 3 can leap over point 2. They are the cases of high-amplitude oscillations which may trigger outbreaks. First explanation of these effects was suggested in *Isaev and Khlebopros* [1977] and more details can be found in *Isaev et al.* [1984; 2001], and *Bazykin et al.* [1997].

Note that some species have adapted in a way that their population outbreaks spare their host ecosystem. For example, some pests of conifers in the Swiss Alps eat only the lower half of needles and ignore the cruder upper half [*Schwerdtfeger*, 1952; 1956]. Thus, trees can survive and grow new needles. Several species of this kind have been found in Russia as well and must exist elsewhere in forest lands. This feeding pattern is good for both the forest and the pest as the parasite is interested in preserving its food resource. In further chapters we discuss other examples of recoverable and unrecoverable resources in relation to the economic activity of man.

Classification of population processes

The straightforward approach we described is not immediately applicable to population processes in general which are as a rule far more complicated. Namely, generations not always take turns at regular intervals, yearly as in insects, or every seven years as in salmon, or any. Moreover, generations can overlap and coexist. The approach works only provided that the following conditions are satisfied:

1. All individuals of the species reproduce simultaneously at regular time intervals.
2. Every former generation dies out after its reproduction phase (which is shorter than the interphase) and before the following reproduction phase of the following generation.

Under these conditions, all possible population processes fit a comprehensive classification. This classification was suggested for the first time in *Isaev and Khlebopros* [1984] and developed in *Isaev et al.* [2001].

There are six possible patterns of population behavior for the cases when the phase curve crosses the bisector at three points (except zero where the population does not exist).

(1) A single stable equilibrium point of the population, as in Fig. 8.

(2) A single unstable equilibrium point, as in Fig. 10 (or Fig. 11), with quasi-random oscillations around it.

(3) Two stable equilibrium points and one unstable equilibrium point, a "point of escape", as in Fig. 12 (or Fig. 17).

(4) One stable equilibrium point, one unstable equilibrium point with quasi-random oscillations, and one unstable escape point, as in Fig. 18, the population at the first unstable point being smaller than at the second one.

(5) One stable equilibrium point, one unstable equilibrium point with quasi-random oscillations, and one unstable escape point, as in Fig. 19, the population of the first unstable point being greater than at the second one.

(6) Two unstable equilibrium points with quasi-random oscillations and one escape point, as in Fig. 20.

The phase curves with two equilibrium points are obviously unrealistic: once got to the right of the unstable point, the population would grow infinitely. The same holds true for any even number of equilibrium points. On the other hand, any odd number of equilibrium points would be theoretically possible, though the known populations have no more than three such points. Thus, our classification appears to be exhaustive for the intended biological applications.

Note that outbreaks of forest insects acquire specific features in the time of climate change like one we might now witness. Sudden warming or cooling make forests vulnerable to pest attacks, especially the forests at habitat borders, e.g. at the forest-steppe interface. On the one hand, it is because trees become exposed to conditions they were not selected for, and their pest resistance weakens. On the other hand, climate change can be bearable for adult trees but fatal for young growth, which can hinder forest recovery. As a result, the outbreaks that would have at most destroyed a part of the forest in the former climatic conditions completely replace it by a steppe ecosystem in a new climate. Climate events can even trigger pest outbreaks in territories where they were unknown for thousands of years. The climate-dependent behavior of outbreaks is explainable in terms of cumulative catastrophes often invoked in describing natural hazards.

Thus we investigated the interaction of pest insects with a forest, their habitat, and the related microcatastrophes of sudden population outbreaks which can either change drastically the state of the forest or reduce its area, or even cause different degrees of deforestation. Although apparently confusing, the behavior of the 'pests-forest' system turns out to follow quite simple laws and the outbreaks of pests develop according to a limited number of scenarios.

We arrived at this result using the method of phase portraits, a phenomenological modeling approach based on empirical input data. Phase portraits are either one-dimensional or two-dimensional images, corresponding to the number of investigated parameters of a system. In the one-dimensional case, they are plotted in the phase planes (x_n, x_{n+1}). Phase portraits in the plane (x_n, x_{n+1}) — applied in this chapter and in some other chapters below — show discrete changes in some system parameter measured at two successive moments of time. The phase plane $(x, \dot{x})$ used in some of the successive chapters displays processes that change continuously, and these portraits (or phase curves) describe the rate of changes. Two-dimensional phase portraits represent the dynamics of two parameters of a system shown along the two respective coordinate axes. In the case of insects, a two-dimensional system can include, for example, the dynamics of pests and of their natural enemies [*Volterra*, 1931; *Kolmogoroff*, 1937].

Phase portraits can be narrow or broad. In narrow phase portraits (very narrow ones can be approximated by a curve), the recorded change in system parameters is defined uniquely by their previous value (for insects it means that a change in the density of a population is defined only by its original numbers). Systems described by narrow phase portraits move along their phase curves and stop at equilibrium points: virtually forever at points of stable equilibrium and for a while at points of unstable equilibrium; the time a system can stay at an equilibrium point corresponds to the time scale of its evolution. The effect of other factors, often quite numerous, is allowed by phase portrait broadening. In the case of insects, a broad phase portrait can image the density of a population controlled by several regulating mechanisms rather than its original numbers: numbers of predators, food supply, etc. Then, a system moves within the limits of its broad phase portrait but remains anchored on the equilibrium points (see Chapter 4), which are the characteristic points of the phase portrait and invariably define the few possible scenarios of the system dynamics. This key feature of phase portraits is applied in this and in the following chapters to predict the dynamics of different natural and cultural systems.

REFERENCES

Baltensweiler W., 1964. Zeiraphera griceana Hubner (Lepidoptera, Torticedae) in the European Alps. A contribution of the problem of cycles, *Can. Entomol.*, 96(5), 792–800.

Baltensweiler W., 1970. The relevance of changes in the composition of larch bud moth populations for the dynamics of its numbers, *Proc. Adv. Study Inst. Dynamics Number Popul.*, Oosterbeek, 208–219.

Baltensweiler W., Benz G., Boney P., and Delucci V., 1977. Dynamics of larch bud moth population, *Ann. Rev. Entomol.*, 22, 79–100.

Bazikin A.D., Berezovskaya F.S., Isaev A.S., and Khlebopros R.G., 1997. Dynamics of forest insect density: bifurcation approach, *Theoretical Biol.*, 186, 267–278.

Berryman A.A., Stenseth N., and Isaev A.S., 1987. Natural regulation in herbivorous forest insect populations, *Oecologia*, 71, 174–184, Berlin.

Isaev A.S. and Khlebopros R.G., 1973. The stability principle in population dynamics of forest insects, *Doklady AN SSSR*, 208(1), 225–228.

Isaev A.S. and Khlebopros R.G., 1977. Delay effects as regulatory mechanisms of forest insect populations, *Doklady AN SSSR*, 232(6), 1448–1451.

Isaev A.S., Khlebopros R.G., Nedorezov L.V., Kondakov Yu.P., and Kiselev V.V., 1984. *The dynamics of forest insect populations*, Nauka, Novosibirsk, 253 pp. (in Russian).

Isaev A.S., Khlebopros R.G., Nedorezov L.V., Kondakov Yu.P., Kiselev V.V. and Sukhovolsky V.G., 2001. *The population dynamics of forest insects*, Nauka, Moscow, 374 pp. (in Russian).

Kolmogoroff A.N., 1937. Sulla theoria di Volterra della lutta per l'esistenza, *Giornale dell'Istituto Italiano delgi Attuari*, 7, 74–80.

May R.M., 1971. Stability in multispecies community models, *Math. Biosci.*, 12, 59–79.

May R.M., 1971. *Stability and complexity in model ecosystems*, Princeton, Princeton University Press, New York, 369 pp.

Population Ecology: Fighting a Forest Pest, 1973. News and Views: Our Soviet Correspondent, *Nature,* 241 (23 February), 504–504.

Schwerdtfeger F., 1952. Untersuchungen über der eisen Bestand von Kiefernspannern (Bupalus piniarius L.), Forleute (Panolis flammea Schiff.) undc Kiefernswarmer (Hylobicus pinastri L.), *Z. Für Angew. Entomol.*, 34(2), 216–283.

Schwerdtfeger F., 1956. Zum Begriff der Populationsdynamik, *Beitr. Entomol.*, 6(5–6), 461–464.

Volterra V., 1931. *Leçons sur la théorie mathématique de la lutte pour la vie*, Gautiers-Villars, Paris, 214 pp.

Chapter 2

Cancer as a Catastrophe in Organisms

The world of living creatures includes two unequal groups of single-celled and multi-celled forms. Single-celled forms, which appeared much earlier than multi-celled ones, are quite numerous and have evolved very far. They have achieved great perfection and adapted to very different, and even extreme, habitats. Their main distinction is that they reproduce by means of division after achieving a certain size in their life cycle. Multi-celled, or metazoan, organisms include all mammals and, hence, man. In the course of evolution, cells in metazoans learnt to serve the organism as a whole. In the advanced forms, such as mammals, single cells are obedient to the organism much more than people of any king or a dictator have ever been obedient to their sovereign or even to the interests of their country. Single-celled organisms, on the contrary, are used to live by themselves and tend to divide unlimitedly. Some of them developed the ability to penetrate into multi-celled organisms and use them as a feeding environment by reproducing there in an uncontrolled way and thus damaging or even killing their hosts. This mechanism is responsible for all infectious diseases. That is why multi-celled organisms developed the immune system to protect themselves from "wild" invaders by means of phages — cells that eat the strangers, and killers — cells that kill them.

However, there is a special disease, cancer, which has a different mechanism even though it likewise stems from the conflict between metazoans and single-celled organisms. Cells in the former — and hence in humans — fulfill numerous functions, because they live and do in a way to be useful to the whole organism being involved in biochemical and biophysical processes and producing certain substances necessary for the organs to exchange with each other and achieve coordinated work. In critical states, say, when the vital organs of brain or heart are in danger, the cells less important for the activity of a multi-celled organism are even ready to sacrifice their lives and become subject to lysis: they die and give up their contents to the cells of the indispensable organs. However, even the most obedient cells are occasionally "spoiled" by unfavorable external effects such as hard radiation, high temperature, chemical or virus attacks. These effects can transform any sound cell into a cancer cell by pushing it back to the primitive pre-metazoan state. Thus sound cells get "wild" and behave like single cells using the organism for their own needs. Normally an organism contains very few spoiled cells relative to the total number of its cells, and the immune system as a rule kills them before they have time to cause any damage. Another adaptation of higher metazoans against this kind of malignant degeneration is that cells normally have some "safety margin" to resist getting wild. For instance, human cells become malignant after ten injuries in the average. This safety in humans is needed to maintain survival for the relatively long life span and is much more solid than, say, in mice who live much less and whose cells run wild after just three strong disturbances. Anyway, having reached some critical level of injury, cells become wild and use the organism in the same way as any parasite uses its host feeding upon it. They set into active reproduction and eventually kill their feeder. The process is similar to insect outbreaks discussed in Chapter 1: during an outbreak cancer cells destroy their host in the same way as pests destroy a forest. Therefore, cancer is actually an outbreak of cancer cells in a multi-celled organism and as such is naturally described by basically the same mathematical model as insect outbreaks [*Bazykin et al.*, 1997].

Today's methods of fighting cancer in the Western medical practice mostly aim at killing cancer cells, most often at the stage of division when they are especially vulnerable. The existing ways of chemotherapy, designed to act at that stage, use the special property of cancer cells to divide continuously unlike many ordinary cells that are mainly responsible for some function — work as a muscle or a liver cell or generate chemical and biochemical substances — and reproduce occasionally. Yet, there are other cells designed primarily for continuous division and reproduction, such as hemopoietic cells in the bone marrow which divide constantly to produce red and white blood cells, or cells inside the stomach or nasopharynx, etc. Naturally the drugs used to kill dividing cancer cells kill those cells as well and thereby upset the work of the hematopoietic system, the stomach, the respiratory tracts and, consequently, the whole immune system, and the organism in general. Other ways to fight cancer such as X-rays are likewise double-edged as they kill cancer cells and their neighbors in tissues, leaving aside the higher risk that radiation gives rise to new cancer cells. Therefore, it is not obvious which is the best choice to cause less damage to an organism than to cancer cells.

All these problems suggest that measures to prevent cancer are as vital as treatment. For this it is important to estimate the contributions from different external factors. The common approach is to inject cancer cells to test animals and watch their response to chemotherapy or X-rays. Although this approach is useful to assess the *therapeutic* effect of external factors, it says nothing about their impact on the cancer *risk*. For example, the numerous attempts through the recent three or four decades to investigate the role of nutrition on the course of cancer have shown no statistically significant dependence. Control animals died, say, on the 14th day, and the test animals died a few hours later on the same day. Yet, recent experiments of scientists from Krasnoyarsk did reveal a significant correlation between the feeding pattern and the cancer risk. The meaning of these experiments can be illustrated by a simple mathematical model.

Below we suggest a mathematical model of cancer, for both its diffuse form and the form of solid tumors. The model makes it possible to quite easily assess the contributions of different external effects to the cancer risk and can thus be useful in cancer prophylaxis.

Diffuse cancer

First we consider the diffuse cancer caused, for example, by Ehrlich cancer cells dispersed throughout an organism without forming compact colonies. Figure 1 shows the general view of curves that describe the reproduction rate V of cancer cells as a function of their concentration x. Each curve corresponds to a certain state of the immune system and represents the total effect of its interaction with a number of carcinogenic factors. Note that zero concentration of cancer cells is associated with a nonzero reproduction rate. In this case "reproduction" corresponds rather to spontaneous onset of cancer cells due to various external and internal effects than to the true reproduction by division. These cells are inoffensive because the immune system relatively easily copes with them and kills them before they can reproduce largely enough. The stage when the immune system is able to suppress the proliferation of cancer cells is recorded by the first falling interval of the reproduction curves. However, as the concentration of cancer cells grows, the immune system becomes no longer able to resist without additional resources which take time to set into action. This effect, called time-delay, accounts for the rising interval of the reproduction curves when the immune response "has no time" to catch up the proliferation of cancers cells. However, at

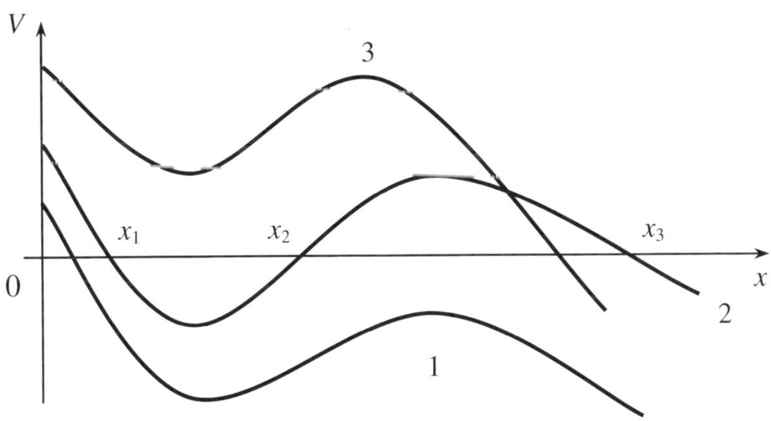

Fig. 1. Reproduction rates of cancer cells V as a function of their local concentration x. Curve 1 = cancer is fully suppressed; Curve 2 = cancer develops after the concentration reaches x_2; Curve 3 = cancer develops even from a single malignant cell and invades the whole organism.

some time, the immune system reaches its full power to fight the further growth of cancer cells. It first slows down and then inhibits their reproduction, which corresponds to the second falling interval of the curves.

Curve 1 crosses the x-axis at a single point and represents the case of a very strong immune system which fully suppresses the onset of cancer. However, numerous factors, including the evolving cancer, weaken the organism and the immune system sooner or later becomes unable to cope with it. Thus, the reproduction curve rises above the x-axis to make a single equilibrium point where cancer cells spread all over the organism and any isolated cell can reproduce to give rise to abundant progeny (see curve 3). Curve 2 crosses the x-axis at three points and corresponds to a rather common case of an immune system unable to provide complete suppression when cancer cells form and develop spontaneously. The point x_1 is stable equilibrium and, like the case of curve 1, corresponds to spontaneous onset of cancer cells in small concentrations inoffensive for the organism. The point x_2 is unstable equilibrium. The immune system suppresses the reproduction of cancer cells until they reach the concentration x_2 and pushes the concentration back to x_1. However, the concentration sets into an ever more rapid growth and reaches the third intersection point x_3 if it leaps over the threshold and happens to exceed x_2 for some reasons. This point is likewise a stable equilibrium corresponding to a very high concentration of cancer cells which exhaust the immune system, invade the whole organism, and eventually kill it. In fact, the power of immune system changes in the course of the disease, and the dynamics of cancer is actually imaged by a series of reproduction curves corresponding to the state of the immune system at each moment of time, as if the reproduction curve were sliding continuously up and down with time. However, curve 2 can account for most of the studied effects as the variations become significant mainly at very advanced stages of cancer which are beyond our consideration.

The cancer risk is evidently associated with the position of the second intersection point x_2. In fact, shifting it to the right to twice higher values means many times lower cancer risk. Of course, each individual has its own phase curve more or less different from those of others. Plotted on the same phase plane, the curves of many organisms taken together give a pattern shown in Fig. 2. One can easily see that the threshold points x_2 are confined within some interval G_0. Thus the pattern splits into three domains: G_0 and the domains on its left (G_1) and on its right (G_2). According to the theory, the cancer risk should approach zero within G_1, and unity within G_2 and takes all values between the two within G_0.

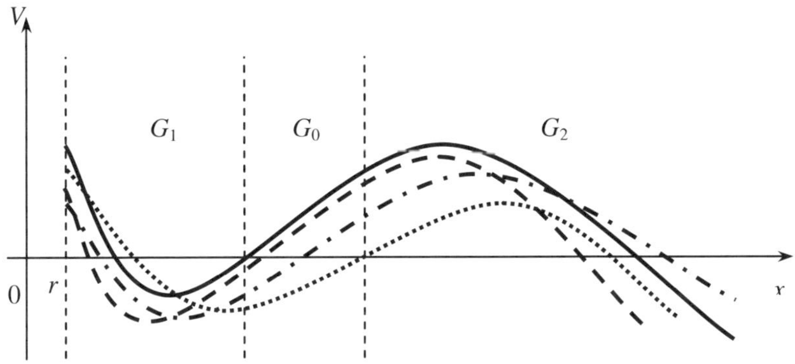

Fig. 2. Distribution of reproduction curves for different individuals and characteristic points to divide the whole plane into three domains of cancer development.

The theory was checked against statistical experiments on test animals (mice) run lately in Krasnoyarsk. Mice were exposed to different doses of Ehrlich cancer cells injected into their abdomen. Rather high doses of about 100,000 cells killed all mice in 15 or 16 days. Lower doses increased the survival time for just one or two days though the variations in the amount of injected cells reached ten times. Some mice survived only after the dose was reduced to about 6,000 cells. The dose of about 4,000 cells was fatal for half of the animals, and doses below 2,500 cells caused almost no mortality. See the mortality rate plotted against the injected dose as solid line in Fig. 3. As we expected, the mortality approached zero and unity at G_1 and G_2, respectively, and varied from zero to unity within the domain G_0 of threshold points corresponding to doses within the range of 2,000 to 6,000 cancer cells.

With these results, it became possible to investigate the effect of nutrition on cancer risk in another experiment[a]. The mice were divided

[a] The idea to apply the experiment as a check of the effect of nutrition on cancer risk belongs to Dr. Soukhovolsky.

into three groups: the control group kept on the usual diet and two test groups of underfed and overfed animals, respectively. Then animals in the three groups were exposed to cancer by injections of cancer cells at a dose of 4,000 which in normal conditions corresponds to 50% survival. The control group showed exactly that death rate but both test groups gave a significantly higher death rate. In terms of our model it means that their threshold point x_2 and, hence, their mortality curve moved to the left (see the dashed-line curve in Fig. 3).

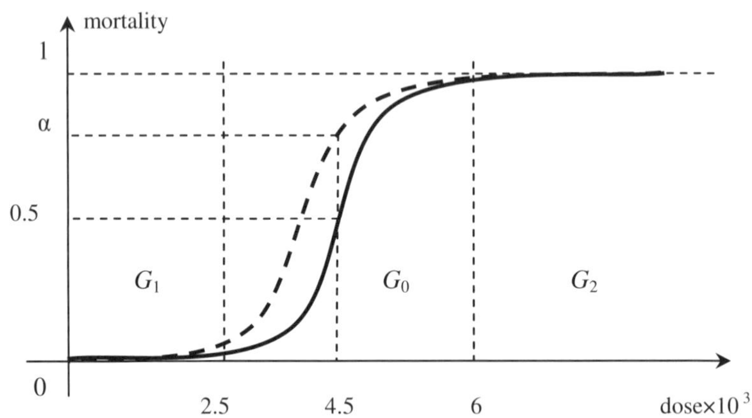

Fig. 3. Mortality rate as a function of injected dose of cancer cells. Solid line = control group; dashed line = test groups of underfed and overfed animals.

Therefore, the cancer risk depends, to a certain extent, on diet, and both malnutrition and overeating increase the risk. Thus, with the only regard to this effect, both advanced and backward economies where people have higher possibilities to overeat or suffer from malnutrition, respectively, may run the same risk of cancer proliferation.

Of course, dietary habits by no means make the only factor of cancer risk. The same method can be likewise easily applied to test environmental effects, food supplements, impact of electrical or magnetic fields, etc. In all these cases moving the threshold point of the phase curve to the right means that an effect is positive in terms of cancer prophylaxis, and negative otherwise.

Solid tumors

Now we apply the same model to tumors, or solid colonies of cancer cells. Newly formed cancer cells are most often very far dispersed in different parts of an organism but occasionally two or more cells may happen to appear close to each other, or, still more rarely, they may have enough time to divide and thus double their local concentration before the immune system kills them. At this point the interaction between cancer cells and the immune system is added with another process associated with the interaction between cancer cells and their sound neighbors. Numerous experiments through the recent three decades showed that neighbor host cells tend to suppress reproduction of cancer cells by producing special biochemical substances [*Abercombie*, 1979; *Sharovskaya and Chailakhjan*, 1999; *Sharovskaya et al.*, 2001]. The higher the concentration of these substances the more the reproduction of cancer cells is suppressed. Therefore, the reproduction rate of cancer cells strongly depends on the number of host cells nearby or, more precisely, on the ratio between cancer and host cells in an area. A single cancer cell surrounded by host cells behaves like a sound cell. This reminds the behavior of a person with ill inclinations (e.g. to theft or any other asocial behavior) who lives among honest people and never develops his vices for this very reason. Otherwise, a host cell surrounded by cancer cells runs wild and divides at any time just like an ordinary cancer cell, against the needs of the organism. Thus the local concentration of cancer cells among host cells was found to be an essential factor of their reproduction rate.

Cells inside a tumor have a limited access to nutrients and oxygen and thus can neither grow nor reproduce. The closer the cancer cells to the tumor surface the better their food and oxygen supply and the greater their contribution to the population growth. In fact, the surface cells produce the greatest part of new cancer cells.

For the sake of mathematical representation, we consider a simplified model which assumes that only cells on the tumor surface can divide. Of course, this is a very rough approximation, as it is often the case in mathematical modeling where it is crucial to distinguish between features of major and minor importance. We mean only a qualitative description, and the model is expected to represent most of the considered effects to a sufficient accuracy.

The reproduction rate of cancer cells depends on their contacts with the neighbor host cells adjacent to the tumor surface, and, hence, on the local concentration of cancer cells in the surface vicinity. Therefore, their local concentration is controlled by the tumor radius. The greater the radius the higher the ratio of cancer cells at the interface between the tumor and the surrounding tissue. See this relationship in Fig. 4 showing two tumors with the radiuses r_1 and r_2, where $r_1 < r_2$. Five cancer cells have seven host neighbors at the radius r_1 and only six neighbors at r_2, i.e., the local concentration of cancer cells at the tumor surface is 5/12 at r_1 and 5/11 at r_2; the correspondence between x and R is univocal in this simplified case.

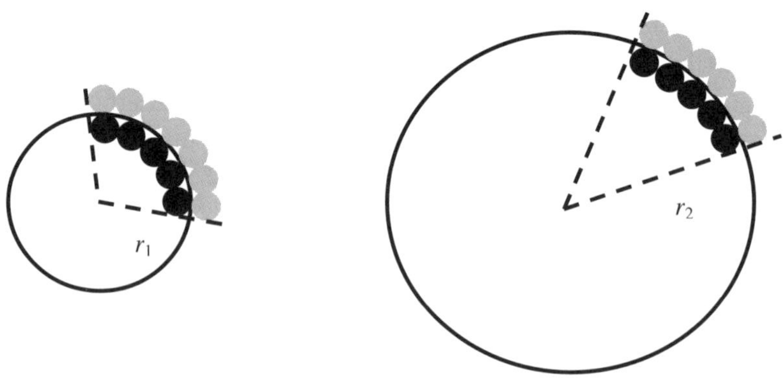

Fig. 4. Local concentration of cancer cells at the tumor surface. Black circles show cancer cells, gray circles show tissue cells.

In a general case it can be found as the ratio $U_1/(U_1 + U_2)$, where U_1 is the volume of a layer of cancer cells that contacts the interface from inside and U_2 is the volume of a layer of tissue cells that contacts the interface from outside. In a two-dimensional case, R and x are related as $R = \pm\, r/((1 - 2x)$, where r is the radius of a single cell. Plus means that $0 < x < 1/2$ and corresponds to a positive curvature radius, and minus means that $1/2 < x < 1$ and corresponds to a negative curvature radius. Note that a tumor with a flat surface has $x = 1/2$ and $R = \infty$. The correspondence between R and x being univocal, the reproduction curve can be plotted as the growth rate W of the tumor as a function of its radius R.

Figure 5 shows the (R, W) reproduction curve for the case when the three equilibrium points are to the left of the point $x = 1/2$. Unlike the case of diffuse cancer, the rising interval of the reproduction curve for solid tumors is accounted for by the effect of local interaction between cancer and host cells rather than by the immune response delay. The response of the immune system has no delay in this case as the characteristic time of tumor growth is at least several months or even years which is largely enough for the immune system to get up its activity.

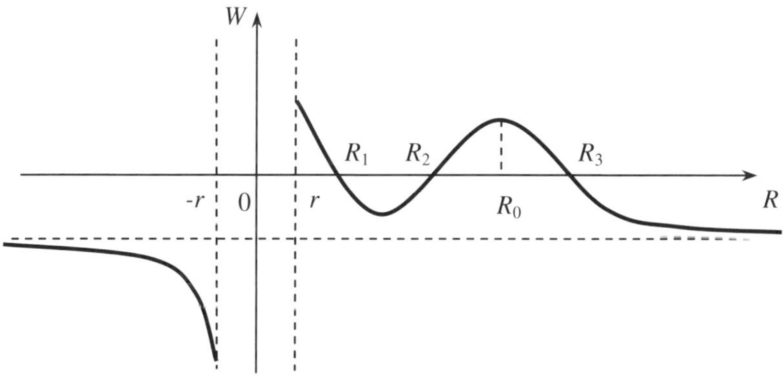

Fig. 5. Reproduction curve in terms of (R,W) where R is the tumor radius and W is its growth rate.

Now we have the equilibrium points R_1, R_2, and R_3 instead of x_1, x_2, and x_3, where $R_i = R(x_i)$, $i = 1, 2, 3$. The point R_1 is the minimum stable equilibrium radius of spherical tumors which is normally within a few micrometers. Tumors of this size can stay for years, disappearing and reappearing, without making any harm to the organism. Yet, if the tumor radius happens to exceed the critical value R_2 for some reasons, the tumor begins to grow until it reaches R_3. The shorter the spacing between R_1 and R_2 the higher the cancer risk. The disease is almost improbable if the two points are spaced rather far but the risk increases greatly if R_1 and

R_2 approach each other as cancer cells are more aggressive or the immune system is weaker.

The reproduction curve allows qualitative description of the effects associated with tumor growth at different stages. Within the interval between R_2 and R_0, where R_0 is the tumor radius corresponding to the highest growth rate, a spherical tumor is stable against minor casual changes in the surface curvature. For example, when the sphere stretches into an ellipsoid, the areas with a smaller curvature radius tend to grow

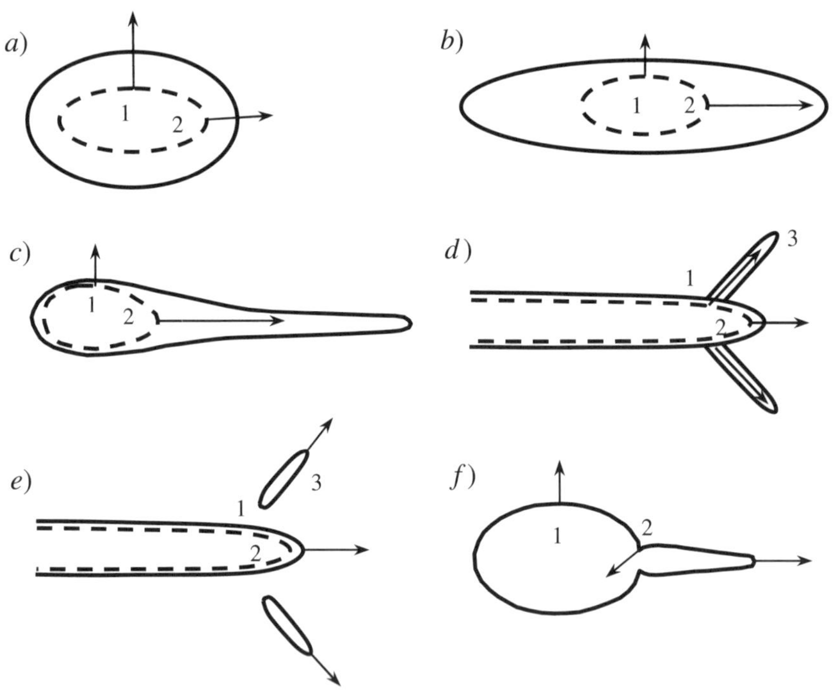

Fig. 6. Growth and formation of tumor at minor deformations. Dashed lines show a deformed tumor, solid lines show tumor shape some time after deformation. Arrows show the magnitude and direction of tumor growth at the corresponding points. a = The initial tumor has a curvature radius between R_2 and R_0. b = The initial tumor has a curvature radius between R_0 and R_3. c = Curvature radius on the end of the "ellipsoid" exceeds R_0. d = Curvature radius at the end of the outgrowth is less than R_0. e = Detachment of the outgrowth in a site of negative curvature. f = Detachment of the outgrowth from the main tumor and development of a metastasis.

more slowly while the areas with a larger curvature radius grow faster. Thus the tumor recovers its original spherical shape, whereby bosses and pits that occasionally arise on the tumor surface smooth out (Fig. 6*a*).

At the initial tumor radius between R_0 and R_3, the tumor spherical shape becomes unstable — as the growth rate is inversely proportional to the curvature radius — and the tumor stretches into an ellipsoid (Fig. 6*b*); at the same time its surface becomes rough to resemble water ripples on a windy day. The tumor surface is ripply in all cases from 6*b* through 6*f*, though the ripples are not shown in the figures being hard to image. The further evolution of this tumor can follow several scenarios. The ellipsoid's sharp end remains the site of the fastest growth rate as long as its curvature radius exceeds R_0 and extends on and on transforming the tumor into an elongate outgrowth (Fig. 6*c*). When the curvature radius at the sharp end of the outgrowth decreases to below R_0, the sites of fastest reproduction move along the stem sides to the points of the curvature $R = R_0$ which gives rise to new outgrowths and the tumor acquires a dendritic structure (Fig. 6*d*). Later on the new outgrowths can detach from the main stem and fall in the region of negative curvature corresponding to a negative growth of cancer cells (Fig. 6*e*). The outgrowth can likewise detach and evolve as a separate metastasis if negative curvature develops at the junction of the outgrowth with the main tumor (Fig. 6*f*). Or, it can travel with the bloodstream and give rise to a new tumor elsewhere in the organism. Once the tumor reaches the radius R_3, its growth progresses exclusively by increasing roughness of the tumor surface which develops outgrowths, branches and dendritic structures.

Crossing the tissue interface

Very interesting effects arise in the case when a tumor crosses the interface of two different tissues. For example, it can penetrate from an organ into a blood vessel or into a lymphatic gland, or from a mucous membrane into the underlying tissue. There the tumor growth depends on the relative position of reproduction curves in the two tissues. Suppose a proliferating tumor reaches the interface of tissues with reproduction curves 1 and 2, respectively, where tumor growth begins at R_2 and ends at R_3 in the first tissue and is between R^2 and R^3 in the second tissue; R_0 and R^0 are, respectively, the points of the highest growth rate in the two tissues; R_c is the local curvature radius of the tumor at the point of contact with the interface between tissues 1 and 2. Theoretically there are three basically different possibilities for the relative position of curves 1 and 2 (Fig. 7).

1. *Intervals* $[R_2, R_3]$ *and* $[R^2, R^3]$ *do not intersect* (Fig. 7*a*). In this case the tumor can never cross the tissue interface, from neither direction, because the interval of positive growth rate in one tissue corresponds to that of negative growth rate in the other tissue. Therefore, the tumor stops proliferating after reaching the interface.

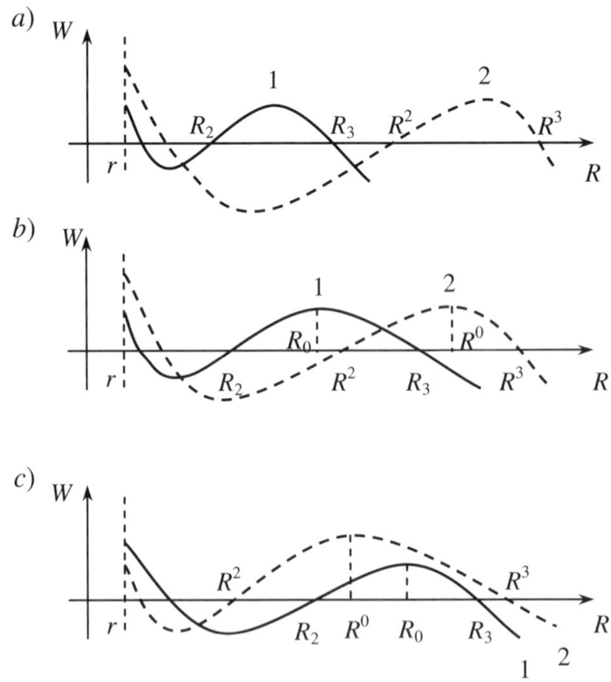

Fig 7. Possible relative positions for the intervals of positive tumor growth rate in tissues 1 (curve 1) and 2 (curve 2). a = intervals $[R_2, R_3]$ and $[R^2, R^3]$ do not intersect; b = intervals $[R_2, R_3]$ and $[R^2, R^3]$ intersect and overlap at $[R^2, R_3]$; c = interval $[R_2, R_3]$ is included inside interval $[R^2, R^3]$.

2. *Intervals* $[R_2, R_3]$ *and* $[R^2, R^3]$ *intersect to overlap within* $[R^2, R_3]$ (Fig. 7*b*). Then, there are two possibilities:

(*a*) Approaching the interface from the side of tissue 1, the tumor fails to cross the interface unless some conditions are fulfilled. If the local radius R_c is smaller than R^2 at the moment of the contact with tissue 2, the tumor stops proliferating until its radius reaches the value R^2 and only then it starts proliferating inside tissue 2. At $R_c > R^2$, the tumor enters freely into the neighbor tissue. Furthermore, at $R_c < R_0$ and $R_c < R^0$, the tumor keeps its growth unstable after it crosses the interface. Stable growth continues at $R_c > R_0$ and $R_c > R^0$ but gives way to unstable growth at $R_0 < R_c < R^0$ whereby the outgrowth which kept its stable shape in tissue 1 sets into active branching and metastazing once it penetrates tissue 2.

(*b*) Approaching the interface from the side of tissue 2, the tumor never crosses the interface at $R_c > R_3$. However, it crosses the interface and proliferates into the neighbor tissue at $R_c < R_3$. Like the previous case, its stable growth may become unstable and vice versa. Note that in this case the tissue interface can become penetrable for the tumor from one side and impenetrable from the opposite side.

3. *One interval is included into the other* ($R_2 > R^2$ *and* $R_3 < R^3$) (Fig. 7*c*). The tumor can always cross the interface from the side of tissue 1 and penetrate into tissue 2 but it can pass in the opposite direction only at $R_2 < R_c < R_3$.

Some special cases

Below we illustrate the theoretically inferred effects with examples of tumor interaction with the surrounding tissues.

Outgrowth penetrating into lymphatic gland. When an outgrowth meets a lymphatic gland on its way through some tissue, it easily enters inside but is known to stay there for a quite long time before it can get outside. Rather, it splits into numerous tiny pieces to be further dispersed all over the organism with the lymph flow. The process fits cases 2 and 3 of one-sided interface penetrability associated with the stability loss of the spherical growth. Note that there we deal with large outgrowths detached from the main stem rather than with single cells or cell aggregates which likewise can penetrate into the gland and initiate the well known process of metastazing. What we describe here is a different mechanism of metastazing through relatively large, and thus more survivable, pieces of a tumor.

Outgrowth penetrating into a blood or lymphatic vessel. The case is similar when an outgrowth penetrates into a blood or lymphatic vessel. The main difference is in the absence of local interaction with the host cells because the tumor falls into a very low-density medium, but the importance of the immune system increases notably. The general mechanism remains the same (Fig. 8): the outgrowth reaches the vascular wall (solid line); then, depending upon the curvature of its tip, it either penetrates through freely or stops for a while, flattening and increasing its curvature radius (dashed line), or perhaps even overrides the vessel, if the interval of positive growth rate for the vessel tissue is on the right of $x = 1/2$ (dotted line).

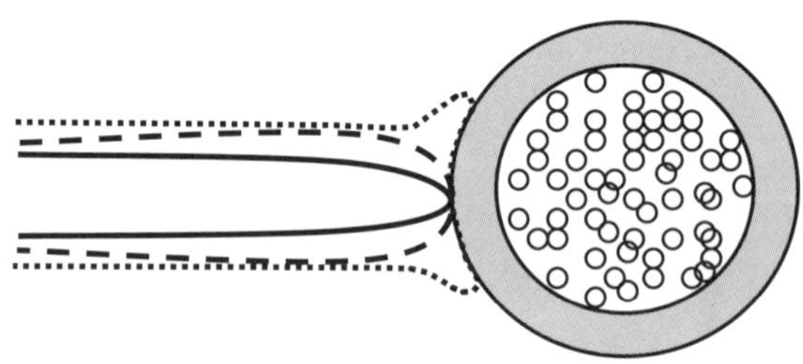

Fig. 8. Successive stages of an outgrowth going through the vessel wall: 1 = outgrowth reaches the wall (solid line); 2 = progress stops until the outgrowth tip acquires the curvature necessary to cross the tumor/vessel interface (dashed line); 3 = the outgrowth tip overrides the vessel wall to acquire negative curvature radius (dotted line). After reaching the required curvature, the outgrowth resumes proliferation, crosses the interface and disperses into small pieces.

Once the outgrowth penetrates through the vessel wall it suddenly enters a very low-density medium strongly different from the one it just left, its growth thus becomes unstable and it breaks into single cells or cell aggregates and becomes disseminated as metastases by the bloodstream throughout the organism.

Stage of vascular tumor growth. Once a tumor reaches some critical size, its inner cells suffer from the lack of oxygen and nutrients and excrete special substances to stimulate the growth and penetration of the neighbor blood vessels inside the tumor. In this case cancer cells can get into the circulatory system right from inside the main tumor. After a vessel penetrates inside the tumor, its wall makes a negative-curvature interface with the surrounding tumor. The vessel becomes penetrable for cancer cells if they reach a local concentration at the interface within the interval of positive growth. There are several ways of reaching this.

Having penetrated inside the tumor, the vessel sets into growth increasing its radius and significantly decreasing the local concentration of cancer cells in the vicinity of the vascular wall. At some point the growing vessel can reach a radius which falls within the interval of the positive growth rate of cancer cells inside the tissue of the vascular wall. Then the tumor penetrates into the vessel by the same mechanisms as an outgrowth does.

The vessel can also change its curvature as it flattens and stretches due to mechanical deformation and acquires an ellipse-shaped cross section with different curvatures at different points of the tumor/vessel interface. Then cancer cells cross the interface at the points where the local concentration x falls into the positive growth interval inside the vascular tissue.

Or, the vessel growing inside the tumor can bend into a loop with different curvatures at different points. In this case, the local concentration x is determined by both the cross and longitudinal sections of the blood vessel.

There is another growth mechanism independent of the curvature of the tumor/vessel interface. Some time after a blood vessel penetrates inside the tumor, the oxygen and nutrients supply to cancer cells becomes sufficient for their environment (the tumor interior) to follow a different reproduction function, with the positive growth interval shifted rather far to the right. Therefore, the cancer cells divide increasing their concentration — and, hence, the internal pressure — so much as to mechanically punch the wall and cross the interface whichever be its curvature.

Prospects: prophylaxis of solid tumors

The dynamics of tumor growth is basically described by the same mathematical model as the proliferation of diffuse cancer, and, therefore, the approach to prevention of diffuse cancer can be expected to work for the prophylaxis of solid tumors as well. This hypothesis can be checked by an experiment similar to the above one, with the only difference that the objective is to produce solid tumors of different radiuses by hypodermic injections of different doses of cancer cells into the belly of test animals and to watch which tumors remain stable and which grow. The threshold radius associated with 50% mortality is found in the same way as the threshold concentration in the case of diffuse cancer. Then the test animals are exposed to various external effects to study their impact on the cancer risk.

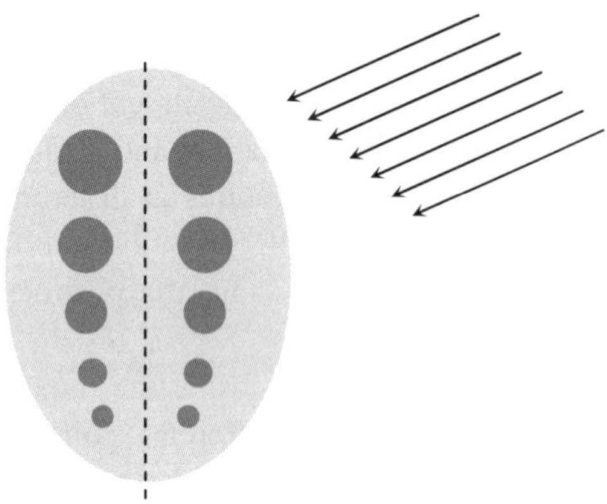

Fig. 9. Sketch of a mouse belly with tumors of different sizes on both sides. One side is exposed to external effects (arrows) and the other remains intact as a check.

The experiment can be still simpler and much cheaper if several tumors of different radiuses are produced on both belly sides (see the sketch in Fig. 9). Then various impacts can be applied to one belly side while the other side is left intact as a check. In both cases the impact of a factor is reflected in changes it causes to the threshold radius.

These experiments are underway in Krasnoyarsk.

Although originally meant as a tool for describing cancer and developing the prophylaxis strategy, our approach can have other applications far beyond the limits of cancerology. Indeed, the approach rests upon a single theoretical postulate, namely the specific behavior of the reproduction curve implying the presence of a threshold point in the cancer evolution. Yet, the same behavior of phase curves showing that a disease develops after the system leaps over the threshold point is found in many other cases besides cancer, primarily in all viral and bacterial infections. Investigation of these diseases on the basis of the approach we suggest is not a very long distance in the future.

REFERENCES

Bazykin, A.D., Berezovskaya, F.S., Isaev, A.S., and Khlebopros, R.G., 1997. Dynamics of forest insect density: bifurcation approach, *J. Theor. Biol*, 186 (3), 267–278.

Abercombie, M., 1979, Contact inhibition and malignancy, *Nature*, 281, 259–262.

Sharovskaya, Yu., and Chailakhjan, L.M., 1999. Local cell interaction and cell growth control, *Doklady Biochemistry*. 366, 80–83.

Sharovskaya, Yu., Gainulina, S.M., Yakusheva, A.A., Kordyukova, L.V., Chailakhjan, L.M., and Aleksandrov, V.B., 2001. Organoid culture of human colon adenocarcinoma as a model of local intercellular interactions, *Biological Science*, 377, 187–190.

Chapter 3

Life and Atmosphere

It is common knowledge that life on the Earth exists due to its atmosphere. The atmospheric gases (dry or dissolved) are vital for living organisms providing their nutrition and respiration. A less obvious but no less essential point is that the atmospheric composition has experienced an extremely strong influence of life. Atmosphere changed the proportion of its constituent gases through the 4.5 billion year history of the planet, and life, once appeared, became an active agent in this process.

The prebiotic atmosphere was largely methanic and contained ammonia, hydrogen sulfide, and carbon dioxide, but no oxygen. The early life may have been chemosynthetic and used the available gases suitable for its existence and development. The amount of those gases reduced progressively with the expansion of biosphere. When the chemosynthetic resources (methane and hydrogen sulfide) became insufficient, life had to choose another energy source. It was the energy of the Sun. The use of sunlight became possible as life invented photosynthesis. Or, according to an alternative hypothesis, photosynthesis may have had its more primitive precursor that appeared prior to chemosynthesis, and the latter arrived as a later adaptation to extreme habitats [*Schnoll*, 1979].

Anyway, land and water life has once become mostly governed by photosynthesis while chemosynthesis remained restricted to few extreme environments of high temperature, acidity, or radiation. In photosynthesis, carbon dioxide extracted from the atmosphere by chlorophyll-bearing green plants, including phytoplankton and green algae, breaks down into carbon and oxygen by the energy of sunlight. Carbon transforms into carbohydrates, which constitute the main building material for all plant tissues, and oxygen is released into air or water. The presence of free atmospheric oxygen is likely an exceptional case in the Universe as, being very reactive, it easily forms oxides or other compounds. Oxygen is real Earth's treasure still not found elsewhere in the Solar system. Plants and animals use it in respiration to gain energy through oxidation of carbon-bearing substances. In fact this is the same process as burning — like that in technological heat-producing devices — but much slower and without fire.

Since the advent of photosynthesis, especially strong impact from biota has been on the carbon dioxide budget, which is the principal subject of the study below.

The content of atmospheric nitrogen is influenced by the nitrogen cycle involved in building proteins. Quite numerous single-celled organisms take nitrogen from air or water actually likewise using the solar energy stored in carbohydrates.

Besides the biochemical cycles, the contents of gases in the atmosphere are controlled by its temperature. For each gas there is a critical temperature at which its molecules approach or even surpass the escape velocity that lets them dissipate in space. That is mainly the reason why atmosphere contains small concentrations of gases with light molecules. The escape temperature for oxygen and nitrogen is not very far from the today's mean global air temperature. This may cause an additional effect on the atmospheric composition during sudden climate changes.

The present dry atmosphere, formed largely due to the activity of biota (flora and fauna), contains about 78% nitrogen (N_2), 21% oxygen (O_2), 0.9% argon (Ar) and minor percentages of other gases, including hydrogen, ozone, methane, nitric oxide, and 0.037% carbon dioxide (CO_2); besides, there is a variable amount of water vapor and water in clouds (0 to 4%, 1% in the average).

Earth's heat budget and surface temperature

The surface temperature of the Earth depends on radiation it receives from the Sun and gives back to space. The received incoming radiation is determined by the temperature of the Sun surface and by the screening and backscattering properties of atmosphere. The Earth is in thermodynamic equilibrium with its cosmic environment, that is, it receives as much energy as it radiates into space. Moreover, the Earth's surface temperature changes rather slowly which fits the principle condition of thermodynamic equilibrium implying constant surface temperature of bodies.

Many bodies which are in thermodynamic equilibrium with their environment radiate electromagnetic waves according to the Planck law and satisfying the Stefan–Boltzmann law. The two laws were rigorously deduced for the so-called black body, an ideal body that absorbs all incoming radiation. The concept of black body is one fundamental idealization of physics like a material point or a homogeneous medium. The Earth is certainly not a black body in this sense as it reflects an essential part of the incoming solar radiation but nevertheless the Planck and Stefan–Botlzmann laws are applicable to terrestrial radiation, as well as to radiation of almost all celestial bodies.

The Planck law implies that the wavelength distribution of the radiated energy is uniquely defined by the temperature of the radiating surface. According to the Planck equation (we do not give it in the explicit form), the temperature of the Sun surface is about 6000K and the greatest part of its radiation is visible light, whereas the outgoing terrestrial radiation, with the Earth's surface temperature about 300K, is mostly infrared (heat). Earth does not emit visible light but glows with backscattered sunlight and is thus seen from space. Thus, although the energy the Earth absorbs is equal to the energy it radiates, the two are of different spectra.

According to the Stefan–Boltzmann law, the total amount of outgoing radiation (let it be W) is proportional to the fourth degree of Kelvin temperature of the radiating surface:

$$W = CT^4,$$

where C is the world constant, the same for all such bodies, and T is Kelvin temperature (measured from the absolute zero of -273°C). Stefan obtained this relationship empirically and Boltzmann then derived it theoretically from the basic thermodynamic principles.

The global mean surface temperature is now of the order of 300K. Its changes can be predicted knowing the amount of additional energy the Earth absorbs. The increment of absorbed energy, in turn, is easy to estimate from increase in the backscattering capacity of the atmosphere related to the release of the known amount of greenhouses gases. At the temperature T' the radiation W' is

$$W' = CT'^4,$$

Dividing this relationship by the previous one gives

$$\frac{W'}{W} = \left(\frac{T'}{T}\right)^4$$

or, substituting $T' = T + \Delta T$ and $W' = W + \Delta W$,

$$1 + \frac{\Delta W}{W} = \left(1 + \frac{\Delta T}{T}\right)^4.$$

If the temperature change ΔT is low relative to T, one can calculate the power in the right-hand side and hold only the first power of the small value $\Delta T/T$; then

$$\frac{\Delta W}{W} = 4\frac{\Delta T}{T}.$$

If the Earth's radiation changes, whichever be the cause, for 1% or $\Delta W/W = 0.01$, its surface temperature will change as $\Delta T = 0.75°$ (from the previous equation $\Delta T/T = 0.0025$, assuming $T = 300K$), i.e., about one degree centigrade.

The incoming solar and outgoing terrestrial radiation goes through the atmosphere. Nitrogen (N_2), oxygen (O_2) and argon (Ar), the main air components, are transparent to light and heat[a] because their molecules are very small relative to the infrared wavelength. If these gases had been the only constituents of the atmosphere, radiation would have freely reach the surface, but it is absorbed by large molecules of carbon dioxide (CO_2), methane (CH_4), and some other so-called greenhouse gases. The greatest portion of heat is screened by water vapor; the second important portion screened by carbon dioxide is due to the presence of photosynthetic biota. In spite of their very low contents, the greenhouse gases and clouds uptake almost 90% of the outgoing terrestrial long-wavelength radiation and turn it back to the surface. Eventually, some part of heat penetrates into space after repeated backscattering by gas molecules but an essential portion returns. This portion can be estimated from the known optical properties of the greenhouse gases.

Earth's biota and a new stable temperature point

We begin our investigation of the life-atmosphere interaction with modeling the effect of atmosphere on the Earth's surface temperature. First, we assume, based on ample theoretical and experimental evidence, that the total incoming solar radiation is constant and look into variations due to the atmospheric screen.

In the starting simplest model we exclude the biotic component. This model is realistic for a prebiotic Earth and can be used to predict the temperature conditions at the point where photosynthetic life began.

With the assumptions of constant solar radiation and absence of life, the problem is simplified and the atmospheric balance becomes amenable to straightforward mathematical modeling using the phase portrait approach (see Chapter 1). The simplest phase portrait that

---

[a] The minor contribution of UV radiation to the total budget is left beyond the scope of the modeling below. For more details of UV radiation see Chapter 5.

simulates the temperature behavior of the Earth's surface and the surrounding air, obtained from the respective differential equation, is shown in Fig. 1 (T_n is the temperature in the n-th year).

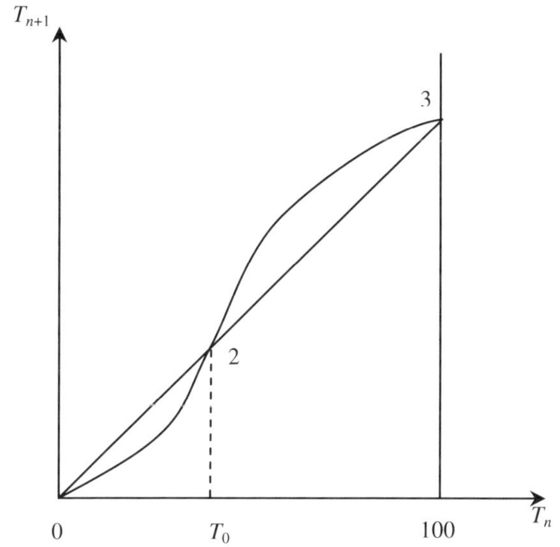

Fig. 1. Physical stability of possible Earth's climates in the absence of biota.

The phase curve has two stable equilibrium points (1 and 3) and a point of unstable equilibrium (2). Point 1 is the ice point and can represent a cold planet such as Mars, and 3 is the water boiling point corresponding to the conditions of a hot planet such as Venus. Both cases imply no liquid water and, hence, life cannot exist in the forms we know it. The region of point 2 corresponds approximately to the temperature range suitable for life, i.e., above 0°C and below 100°C.

The two stable points (1 and 3) being unsuitable for life, we focus on the temperature T_0 at unstable point 2. Scientists from Tomsk [*Zakharov et al.*, 1992] were the first to understand the meaning of this point: it is near point 2 that life could begin. Despite the instability of this point, life upheld in its vicinity long enough to withdraw the appropriate amount of atmospheric greenhouse gases (first methane and then especially carbon dioxide) to reduce the greenhouse effect and cool the Earth's surface.

Thus the phase curve of Fig. 1 went down. This was impossible before the advent of photosynthesis when the content of carbon dioxide was controlled by abiotic agents. Shifted down, the phase curve sagged below the bisector (Fig. 2) to yield a new stable point (*a*)[b]. Thereby life created and occupied a new thermodynamically stable point (*a*) near the unstable point (2); besides, it produced another unstable point, the point *e*. The stability of the new system fits the temperature range between 2 and *e*. It means that the system can be drawn by evolution to the conditions of a Mars-like frozen planet if shifted to the left of 2 and tend to a Venus-like boiling planet if shifted to the right of *e*. Since it created the two new points (*a* and *e*), life has moved the phase curve progressively lower and broadened the sag, i.e., broadened the temperature range of its sustainable existence.

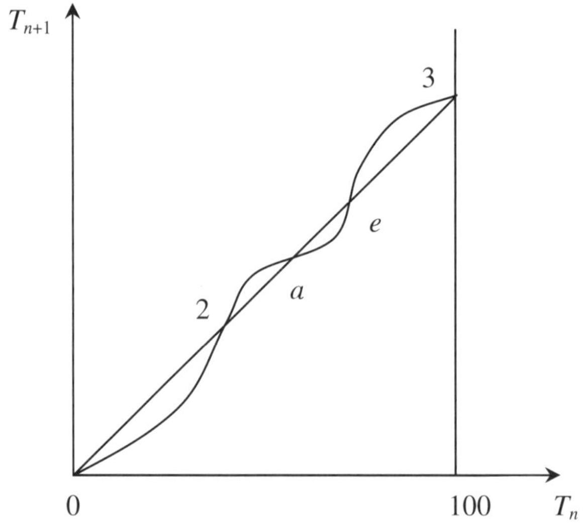

Fig. 2. Phase portrait of the climate system changed under the effect of biota. Arabic numerals (1, 2, 3) denote the equilibrium points of lifeless Earth and letters (*a*, *e*) denote the equilibrium points produced by biota.

[b] In the text and in the figures below we use Arabic numerals (1, 2, 3) to denote the equilibrium points of lifeless Earth and letters (*a*, *b*, *c*, *d*, *e*) to denote the equilibrium points produced by biota.

The vicinity of point 2 in Fig. 2 is not to scale for better visualization. Of course, the effect of biota on climate was at first minor but it grew ever more important as life progressed.

Makar'yeva and Gorshkov [2001] from St. Petersburg developed the remarkable idea of the Tomsk scientists. They obtained a more exact phase curve of Fig. 1 (without effect of biota) which had on the right of point 2 a segment almost parallel to the bisector (see it enlarged in Fig. 3). The key importance of this segment for the life evolution apparently escaped their attention. We suggest a hypothesis to explain it in this book for the first time: It may have been an interval of slowly changing temperature, long enough relative to the Earth's age, for life to have time to install not far from point 2 and to ensure conditions of its further evolution.

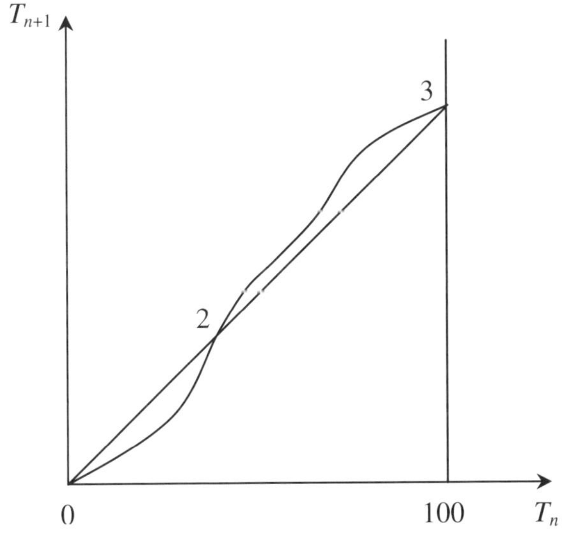

Fig. 3. Enlarged segment on the right of point 2 almost parallel to the bisector corresponding to an interval of slowly changing temperature where life could set up.

That period may have lasted hundreds of million years. We still do not know for sure how did life begin but it obviously would have ceased if the Earth's surface temperature had suddenly risen (Fig. 1). That was another almost incredible chance which permitted the onset of life. Once having appeared, life took a regular control over climate by cycling atmospheric gases and especially carbon dioxide.

Life and carbon dioxide

Carbon dioxide makes only 0.037% of total air composition but is critical for life being the source of carbon for all photosynthesis and an active greenhouse gas, a vigorous absorber and backscatterer of the outgoing infrared terrestrial radiation. Our model can include the greenhouse effect using simple physical ideas that lead to a differential equation where the change rate of temperature is a function of T (Earth's mean global surface temperature) and C (carbon dioxide content). The full derivation can be found in [*Semenov*, in press] and we limit ourselves to imaging the final result in a geometrical form. At the stationary temperature T we have

$$g\,(T,\,C) = 0 \tag{1}$$

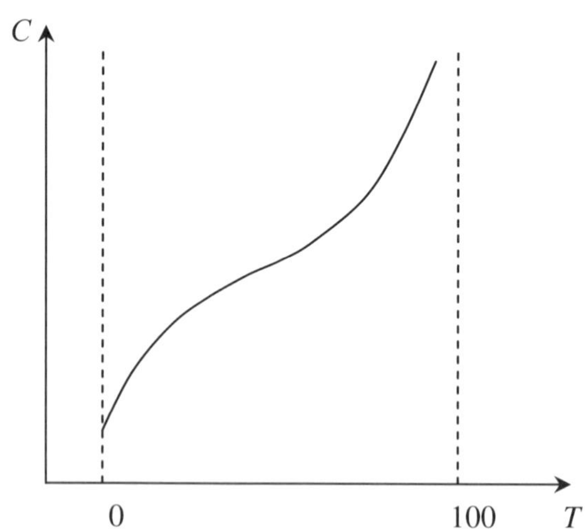

Fig. 4. T-nullcline as a monotone increasing function $C(T)$. Dashed line limits the temperature range suitable for life.

The set of points in the plane (*T*, *C*) that satisfy Eq. (1) is represented by a curve which is called the *T*-nullcline (Fig. 4).

The curve *C*(*T*) is monotone increasing and has a single point of inflexion (Fig. 4). Equation (1) includes the greenhouse effect of water vapor and the present concentration of carbon dioxide. The effect of life is taken into account by Eq. (2) below. Note that the model simulates the *global mean temperature* fitting the range between 0 and 100°C, which does not rule out the presence of polar ice caps, mountain glaciers, and winter snow.

Equation (1) corresponding to the *T*-nullcline in Fig. 4 can be written in the explicit form. It describes the stable level of temperature *T* for the stationary concentration *C*. The zero constant in the right-hand side of (1) is defined by the Earth's physical conditions and is independent of biotic factors. Therefore, the effect of biota on *T* and *C* manifests itself in the motion of the point (*T*, *C*) along the same curve (Fig. 4); if the life state does not change, *T* and *C* do not move. The content of atmospheric carbon C (in carbon dioxide[c]) is controlled by organic and inorganic factors. Most inorganic CO_2 comes with volcanic eruptions and some is released from rocks. This source is assumed to be constant through long geological time unless noted otherwise. Oceanic water likewise releases CO_2 (though absorbs it as well and, possibly, more than releases). The excess carbon the water-living organisms use to build their shells is deposited on the sea floor as calcareous rocks and falls out of the carbon exchange between biosphere and atmosphere. Other life "wastes" dropped out of the carbon cycle include peat, coal, and oil. The total amount of irreversibly withdrawn (buried in the lithosphere) carbon per unit time *f*(*T*, *C*) depends on *T* and *C*. There is reason to believe that this amount is in the first approximation proportional to the biosphere production *P*(*T*, *C*) in biomass units. The existence of *f*(*T*, *C*) shows that the atmosphere-biosphere carbon cycle is open and is determined by evolution. The appearance of free oxygen in the atmosphere and

---

[c] There are other carbon-bearing atmospheric gases that matter, especially methane, but their contents are times lower than CO_2 and including them in the model does not change our further conclusions.

dissolved oxygen in the water made organisms seek for protection from the oxygen attack, i.e., from rapid oxidation. Evolution solved the problem by inventing substances which cannot be recycled because resist oxidation and, hence, drop out of the carbon cycle.

The atmospheric carbon balance is given by

$$w - f(T, C) = 0 \tag{2}$$

where w is inorganic carbon input per unit time.

The function $f(T, C)$ cannot be obtained theoretically and only some qualitative predictions are possible. We assume that $f(T, C)$ is a convex function of its two variables (Fig. 5). This is apparently the simplest possible case. The function of the biosphere production $P(T, C)$, proportional to f according to our assumption, behaves in the same way.

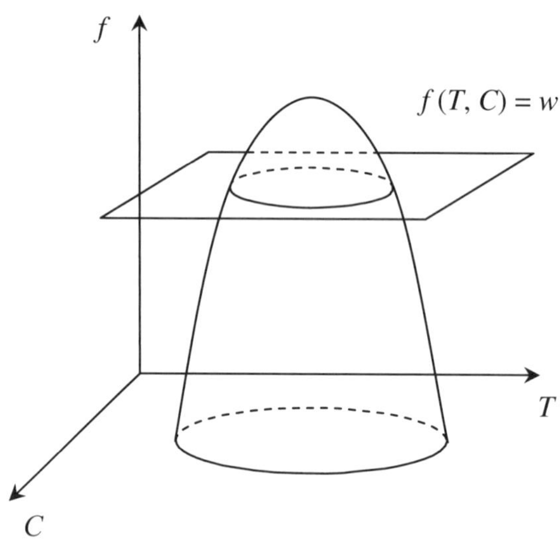

Fig. 5. Total amount of irreversibly withdrawn carbon per unit time $f(T, C)$ depending on T and C.

The convexity of $f(T, C)$ suggests that the curves in the plane (T, C) given by (2) are likewise convex. They are the sections $f = w$ of the surface from Fig. 5 projected onto this plane. Figure 6 shows only the lower portions of the curves which include possible stationary values of

(T, C). We cannot use the curves along their full length as living organisms cannot maintain the stationary state of carbon (control atmospheric carbon) at very high C.

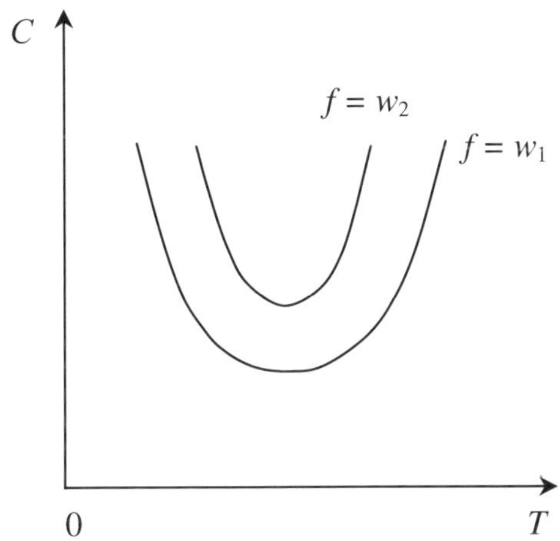

Fig. 6. The lower portions of the convex curves in the plane (T, C) given by (2) which are the sections $f = w$ of the surface from Fig. 5 projected onto (T, C).

At the given form of the function f (proportional to biosphere production), the lower position of the curve corresponds to lower w and vice versa. Therefore, life can exist within a broader temperature range at lower inorganic input of atmospheric carbon. The result is the same at invariable w if the biosphere production increases in the course of evolution (or due to advanced agriculture as it occurs today). Indeed, the surface $f = w$ of Fig. 5 gives way then to a higher one which broadens the temperature range suitable for life.

Stable points of life

The stable states of the system (T, C) should satisfy two conditions that (i) the respective points of the plane (T, C) lie on the T-nullcline of Fig. 4, i.e., be consistent with the Earth's abiotic conditions and (ii) lie on the C-nullcline, i.e., be consistent with the biosphere equilibrium.

Figure 7 shows the simplest case of the two intersected nullclines when the T-nullcline plots a monotone increasing function $C(T)$ and the C-nullcline is truncated to fit the temperature range suitable for life.

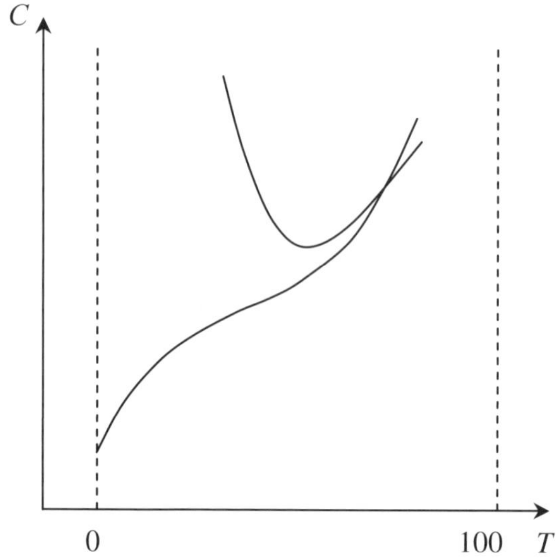

Fig. 7. The simplest case of the two nullclines crossed at a single stable point of the biosphere.

The curves have a single intersection point: an increase in biosphere production (or a decrease in inorganic carbon input w) shifts the intersection point down along the invariable T-nullcline which means a decrease in temperature and in atmospheric carbon dioxide. Thus biota controls the surface temperature and the atmospheric oxygen and carbon dioxide budgets.

Specified behavior of the curve $C(T)$: effect of glaciation

Of special interest is a realistic case of the surface partly covered with ice. This case was studied [*Semenov and Khlebopros*, 2002] using the methods of Poincaré's qualitative theory of differential equations. We can give an idea of the results in geometrical models.

First of all, the phase curve of Fig. 4 turned out to be inexact. The more faithful pattern was obtained in an experimental study by *Pearson and Palmer* [2000] and recorded repeated periods when temperature fall was accompanied by carbon dioxide rise. *Pearson and Palmer* [2000] explained the reversal by a lag in the biota response to climate change. D. Semenov [*Semenov and Khlebopros*, 2002; *Semenov*, in press] hypothesized that the phenomenon revealed by the experiment is not casual and interpreted it in terms of nullclines as a breakdown in the monotone run of the curve $C(T)$. He attributed the change in the nullcline slope when the curve shifts up at low temperature to the appearance of territories covered with ice and snow, or glaciation (see Fig. 8; the original monotone curve is shown by dashed line).

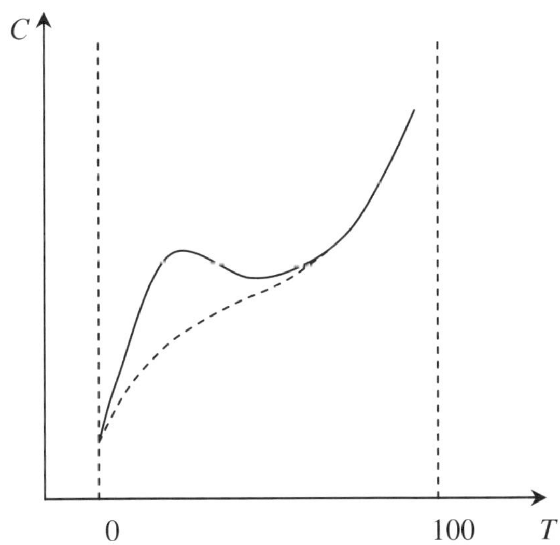

Fig. 8. Breakdown in the monotone run of the increasing function $C(T)$ at low temperature, possibly due to glaciation.

So far we considered the Earth as a "point" system with the constant temperature T. Yet it is actually the global mean temperature. There are estimates that it was presumably about –15°C (or above 100°C) in the absence of life. The existence of life means that temperature should range between 0°C and 100°C. However, there were periods in the Earth's history when some territories were frozen. The presence of perennial ice and/or seasonal snow increases the albedo of these territories, i.e., more solar radiation becomes reflected and less radiation is absorbed. As a result, the global mean temperature falls. At the same time, another consequence of glaciation is that it reduces the biosphere volume as the vegetation cover partly disappears or stops its activity for a while, thereby increasing atmospheric carbon dioxide. Therefore, the breakdown of the monotone behavior of the phase curve allowed a plausible explanation of glaciations [*Semenov*, in press].

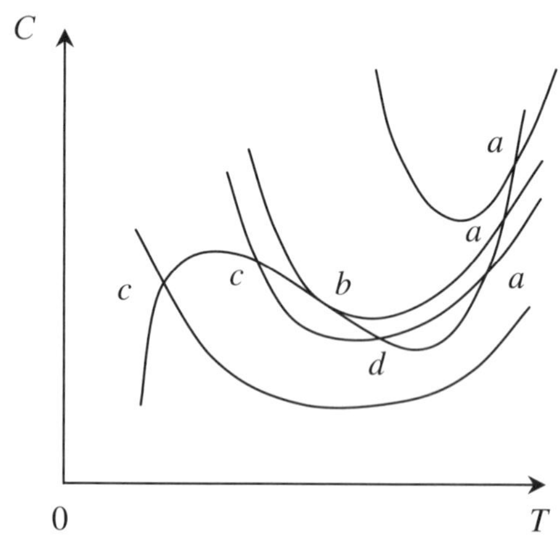

Fig. 9. Two-dimensional phase portrait of global surface temperature under the effect of biota (more exact T-nullcline), in the coordinates of temperature and carbon dioxide, showing intersection of the two nullclines at the points of the biosphere equilibrium.

Global temperature dynamics under the effect of biota

We can interpret the dynamics of the mean global surface temperature under the effect of biota using one-dimensional and two-dimensional phase portraits (see Chapter 1). The two-dimensional phase portrait, in the coordinates of temperature and carbon dioxide (Fig. 9), shows the T and C nullclines crossing at a single stable point a if the C-nullcline is high enough; its descent brings to the appearance of first the tangent point b which holds for a short while and then two additional intersection points c and d. As a result, there occur three possible equilibrium states of the biosphere. Finally, the two points c and d merge on further descent of the C-nullcline, and the system again has a single equilibrium point (c) corresponding to low temperatures, and the phase portrait returns to that of Fig. 2. The point c can fall either on the upgoing or the downgoing branch of the C-nullcline. It is stable in the former case and the mean global temperature changes very slowly for hundreds thousand years. Once it falls on the downgoing branch, the point looses its stability and the global temperature oscillates around it.

Figure 10 shows the one-dimensional phase portrait of temperature dynamics controlled by life evolution in the temperature coordinates (T_n, T_{n+1})[d].

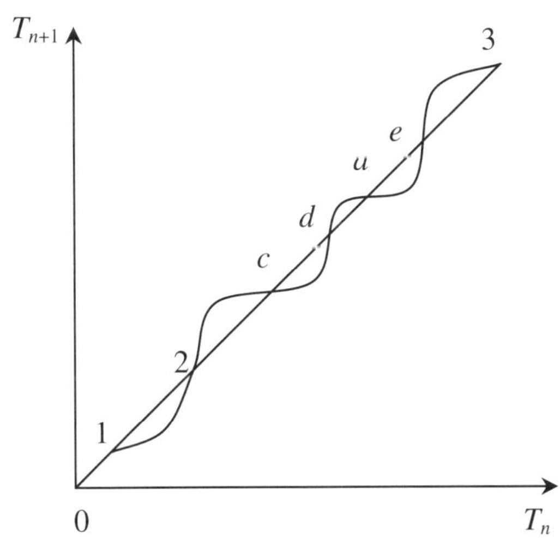

Fig. 10. One-dimensional phase portrait of global surface temperature under the effect of biota, in the temperature coordinates. See text for explanation. Arabic numerals (1, 2, 3) denote the equilibrium points of lifeless Earth and letters (a, c, d, e) denote the equilibrium points produced by biota. The phase curve is rather a qualitative image and the distances between the points are not to scale.

[d] This phase portrait is of the type used in Chapter 1.

The phase curve has seven equilibrium points. As noted above, two extreme points, 1 and 3, correspond to the stable states of a frozen and a boiling lifeless planet, respectively, and are divided by point 2. The onset and evolution of biota produced four more equilibrium points: the point *a* of stable equilibrium, the points *d* and *e* of unstable equilibrium, and the point *c* where equilibrium can be either stable or unstable (Fig. 10)[e]. The point *c* is a "divide" between cold and warm climates, and the point *e* divides the state of a warm climate from that of a hot Earth.

At its beginning life may have been at the only possible stable point *a* (see Fig. 2, 9). It may have been the state of biosphere at the time when photosynthesis (or its precursor) first came into play. In the course of its evolution, photosynthetic biota changed its phase portrait into that of Fig. 9: biosphere remained at the point *a* though another possible equilibrium state at *c* had appeared. Further evolution may have smoothened the curve to make the state *a* disappear and leave a single equilibrium point (*c*) (see the lowermost curve in Fig. 9). The life evolution is recorded in the downward motion of the *C*-nullcline, and its upward motion may be associated with increased inorganic input of carbon dioxide. The interplay of these two basic trends repeatedly moved the temperature setting from *a* to *c* and back.

The existence of the equilibrium points *a* and *c* is confirmed by geological and paleontological evidence of regular alternation of cold and warm periods in the Earth's history. Cold times correspond to the position *c* and warm times to *a*. The warm Earth in *a* should have the global mean surface temperature of the order of +25°C or higher; in this state, there is almost no perennial or seasonal ice and snow, and the subtropic and tropic climate spreads over almost all continents. In *c* the mean temperature is much lower (of the order of +15C°) and latitudinal climates are represented in a broader range of tropic, subtropic, savannah, steppe, boreal forests, and tundra zones which apparently arose successively as far as life evolution progressed. In the cold state *c*, the Earth can be partly covered with ice: Ice exists as polar caps,

---

[e] According to Poincaré's theory, the point *a* always corresponds to stable equilibrium and equilibrium in the point *c* can be either stable or unstable.

mountain glaciers, ice deserts; seasonal snow occurs in the middle latitudes. As *c* looses stability if falls on the downgoing branch of the *C*-nullcline, the oscillations of the global temperature around it correspond to glacial/interglacial cycles of different lengths and strengths with waxing and waning ice sheets. The today's climate fits the conditions of *c*.

As just mentioned, biosphere can move to *c* with a colder climate (and conditions unsuitable for many living species) spontaneously as a result of the biota progress, as it occurs according to the mathematical theory of catastrophes. Or, this shift may be triggered by cosmic events, such as bolid impact. A bolid impact can decrease the surface temperature by dramatically reducing the air transparence for a quite long time (but relatively short on the geological scale). This effect is often invoked to explain the extinction of dinosaurs. The biosphere changes caused by catastrophic events mean a relatively rapid shift of the stable point along the *C*-nullcline. The switch back to warm climate (from *c* to *a*) is not spontaneous but can be triggered by abiotic events that rapidly increase atmospheric CO_2 (*w*), or by man-induced growth of atmospheric carbon dioxide (see Chapter 4).

Thus, not only the Sun and the Earth guided the evolution of life by changing the air temperature and composition but life itself contributed greatly to the formation of the atmosphere and fixed the surface temperature far from the states of a frozen or a boiling planet. Moreover, it was life that created two new equilibrium states of global climate: a warm point with tropic or subtropic conditions almost all over the globe and a cold point with the presence of ice and different latitudinal climate zones.

REFERENCES

Makar'yeva A.M. and Gorshkov V.G., 2001. The greenhouse effect and the stability of the global mean surface temperature, *Doklady Earth. Sci.*, 377(2), 210–14.

Pearson P.N. and Palmer M.R., 2000. Atmospheric carbon dioxide concentrations over the past 60 million years, *Nature*, 406, 695–699.

Schnoll S.E., 1979. *Physical-chemical factors of the biological evolution*, Nauka, Moscow (in Russian).

Semenov D.A. and Khlebopros R.G., 2002. Biophysical aspects of the effect of biosphere on the global climate, *Doklady Earth. Sci.*, 388(2), 230–31.

Semenov D.A., in press. The contribution of biota to the global climate dynamics on the scale of geological time.

Zakharov V.I., Gribanov K.G., Prokop'ev V.E., and Shmelev, 1992. The effect of the 8–3 μcm transparency band on the stability of the Earth' thermal state, *Atomnaya Energiya*, 72(1), 98–102 (in Russian).

Chapter 4

Technosphere-Biosphere Interaction and Global Climate

The air composition has changed through the long Earth's history under the effect of abiotic and biotic factors. Since its beginning, life has been an active agent in the formation of the atmosphere and is largely responsible for its present state. More change to the atmosphere composition — and, hence, to climate — has been due to the activity of man. The human impact grew with the progress in technology. In prehistoric time it consisted in improvement of hunting tools; it caused relatively little harm to nature and had no influence on the atmosphere. The technological pressure on climate and environment has increased dramatically as a result of land use and especially the use of fossil fuels use during the industrial era. The danger has become severe lately and will be aggravated in future if the machinery grows in amount but does not advance towards environment-friendly operation.

Below we investigate the technosphere-biosphere interaction using the methods applied in the previous chapters (see Chapters 1–3), but before we analyze the dynamics of energy use associated with the progress in technology.

Energy use

Nowadays the interaction between civilization and biosphere is determined by the state of technology more than by population or food dynamics. The present technology level in many densely populated countries lags behind that in less populated states. Population growth stimulates and, on the other hand, impedes the technological progress but there is no unique relationship between the two. As for food, the diet and the per capita food intake has not changed much through time as it follows the invariable physiological needs of humans. Thus, the absolute amount of required food grows roughly proportionally to population.

The level of technology can be measured approximately against energy use. By energy use we mean total annual amount of energy produced (and hence used), including the losses due to transportation drawbacks, such as heat loss from power lines which is just energy wasting.

Energy use grew through the historic time and its growth has become exponential since the industrial revolution after 1750. Energy use in prehistoric times was restricted to burning wood and other biomass in open fires and fireplaces and its magnitude was likely proportional to population. Later, in the age of agriculture, energy use (likewise in the form of biomass fuels) grew proportionally. Thus, one can assume that the per capita energy use was roughly invariable before 1700 (Fig. 1 b). The use of fossil fuels and other sources (for example, hydro power) since 1700 has been more or less reliably documented [UNO Energy Statistics Yearbook, 1982; 1992] (see Table 1). The same Table shows the use of biomass fuel according to *Smil, 1994*. The latter estimates are approximate as the household uses remained mostly beyond the statistical accounts.

Table 1. Mean annual energy use, in million tons of oil equivalent.

Year	Fossil fuels	Biomass fuel
1700	4.4	240
1750	13.1	300
1800	31.0	340
1850	64.6	360
1900	427	450
1950	1550	690
2000	10000	690

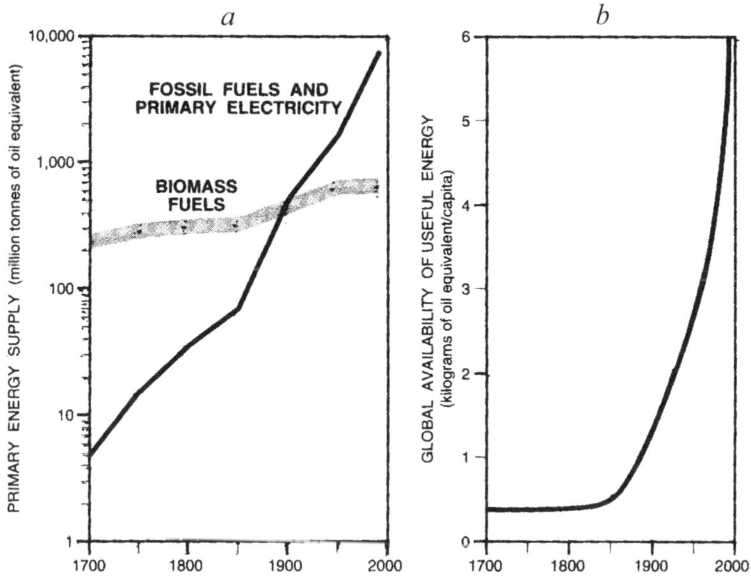

Fig. 1. Global energy supply through 1700–1990. Consumption of fossil fuels (black line, logarithmic scale) and biomass (gray line; its thickness shows possible scatter) (*a*); global availability of useful energy (*b*). Borrowed from *Smil, 1994*.

The energy supply growth in Fig. 1*a* is approximated by a straight line over every fifty years: $y = kx + a$ where k and a are different for different legs. The slope k measures the rate of exponential growth over fifty years.

The growth was especially rapid through 1850–1900 (advance in technology, advent of electricity) and through 1950–2000 (post-war development). The slower growth through 1900–1950 spans the time of the two world wars. The growth rate is especially spectacular in the non-logarithmic curve of Fig. 1*b*.

The worldwide output of fossil fuels surpassed the total supply of biomass energy (gray line in Fig. 1*a*) just before 1900 and has grown more than tenfold since then; the today's biomass consumption is within 6% of the total energy output [*Smil*, 1994].

The growth has slowed down in recent decades changing from exponential to linear (see the more exact plot of Fig. 2), which indicates more slowly increasing demand (saturation) in rich countries which are the major energy consumers. However, one can expect that the third world that has been a modest energy consumer so far would follow the western way of life to cause another sudden rise in global energy use.

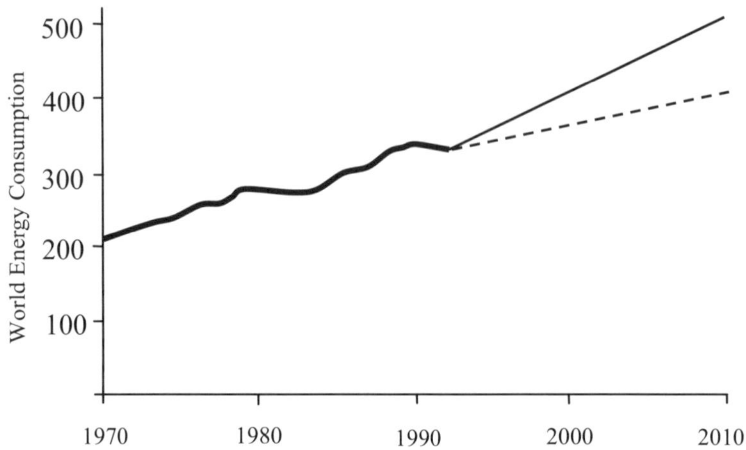

Fig. 2. Total worldwide energy use (after *International energy outlook,* 1995). Heavy solid line is energy use according to historic accounts, thin solid line is predicted maximum, dashed line is predicted minimum.

Worldwide electricity generation has been growing much faster than the supply of fossil fuels [International energy outlook, 1995]. The largest economies have always been the world leading energy producers. Thermal energy still dominates the global output. Alternative power sources (Fig. 3*b*), which are particularly attractive as they cause no air pollution, remain much less important. The share of electricity generated by nuclear power in 1990 roughly equaled the share of hydro power stations and made only one tenth of global output each. Geothermal (underground hot springs) and wind (and tide) power have been marginal sources [*Smil*, 1994].

The future of energy production may lie with solar energy, though it is not shown in Fig. 3 being almost never used for electricity; its application is restricted to local heating.

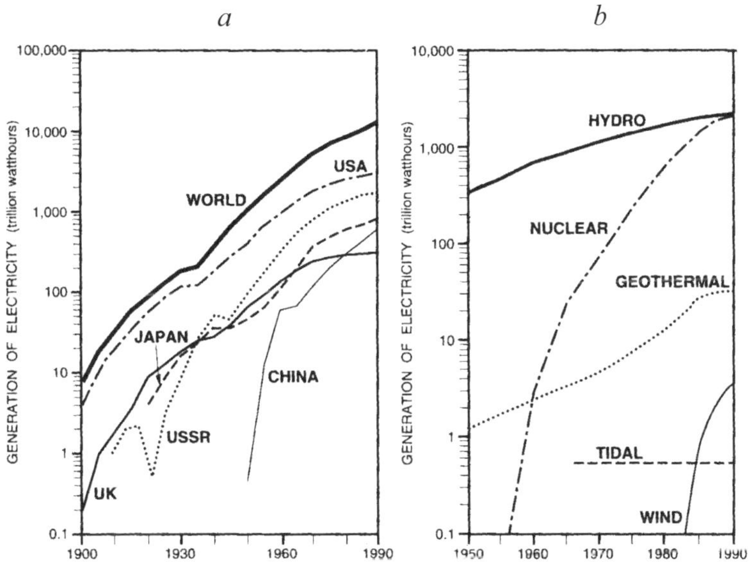

Fig. 3. Electricity generation through 1900–1990, worldwide and by largest economies (*a*) and generation using power sources other than fossil fuels (*b*). Borrowed from *Smil, 1994*.

Below we apply the phase portrait approach (see Chapter 1) to model the dynamics of energy use in the remote past (without statistical support), at present (through recent 300 years, based on statistics), and in the future (predicted). We image the processes only qualitatively and are aware that the plots are not very exact, even for the statistically covered periods. All plots are not to scale.

Let the total global energy output of a current year be K and that of the following year be M. Two successive measurements that give the pair (K, M) are similar to the standard observation from Chapter 1. We assume that M depends uniquely on K (as well as the population density in each following year depends on that in a current year) as energy use represents to a large extent the level of technology which, in turn, controls the future energy demand. Therefore, M is a function of K, or $M = F(K)$. This function is generally increasing as people use ever more energy, at least through the recent 300 years, though the energy use may undergo a decline as well. It may have fallen in the early medieval time in the conditions of a global culture decay or in the beginning of the 18th century in England when the wood demand from the developing metallurgy exceeded its availability and energy consumption may have reduced until coke came into use. Yet, these hypotheses are difficult to check for the lack of statistical data.

Note that the dynamics of energy use differs from the population dynamics addressed in Chapters 1 and 2. Unlike the population dynamics, the continuously growing energy use never returns to any previous level, so that each measured pair (*K, M*) occurs only once. This makes the modeling more difficult as the predictive power of phase portraits stems from repeated situations when each value of *M* can be predicted from *K* (see Chapter 1). However, this is never the case with human activities as no period of history ever repeats. Nevertheless, energy use has its own dynamics, especially if it is globally averaged to cut off local peaks. The simplest phase portrait of energy use corresponds to a single type of fuel, e.g., fire wood. Our ancestors stoked fires and then ovens, and energy use grew gradually till the stable point (point 1 in Fig. 4). Its coordinates represent energy use for very long

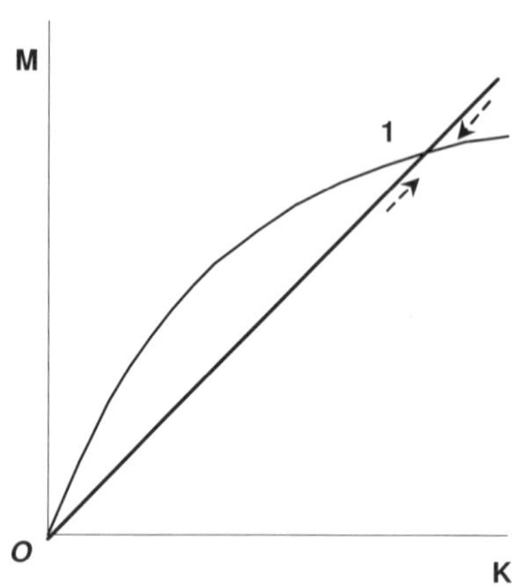

Fig. 4. Simplest phase portrait of energy use based on a single power source, for example, fire wood.

periods of time with minor casual fluctuations. Slow population growth means that the phase curve moves up but its shape remains invariable as the energy demand and the type of fuel does not change. The position of point 1 on the K axis moves up slowly, i.e., energy consumption increased proportionally to population, but the climate balance could be maintained for millennia as the forest and other biomass resources were immeasurably rich against the people's needs.

Wood is a perfect fuel in terms of environment impact provided that the related deforestation is kept up by the natural recovery or plantation of forest. Of course, burning wood releases carbon dioxide, but newly grown trees intake and bind the same amount of carbon dioxide as was contained in the cut and burnt trees, so that the net atmospheric carbon dioxide does not rise (the same holds for any biomass fuel). Other substances released in charring wood are safe for nature and for people. Thus wood had been an ideal fuel until its demand rose dramatically due to industry advance. In the mentioned case of 18th century England, rapid progress of metallurgy on charcoal caused uncompensated deforestation and the price for wood went up. This was one of the first industry-caused environmental disturbances when the natural growth could not make for the cut forests. However, it was the need in wood and not environmental problems that was the main concern at that time. Charcoal soon gave way to black coal which could be produced in abundance in the country, and the industrial revolution received a new impetus.

Without the use of black coal[a] the phase portrait of energy use would remain the same as in Fig. 4, i.e., it would return to point 1 again and again, and bring the metallurgy development to a stop.

If we assume that wood consumption had stabilized by the time when coke came into use, the coal fuel K_c adds to the stable value of wood fuel K_w on the right of point 1 (Fig. 5*a*) and the total energy becomes $K = K_w + K_c$. The total energy of the following year is $M = M_w + M_c$ where M_c is derived from K_c using the phase portrait of coal consumption shown in the same figure with the origin of coordinates in the point (1, 1). The behavior of the coal phase curve must be qualitatively the same as that of the wood curve but the respective arc over the bisector must be much larger as the coal resources are far more abundant and the coal

[a] The idea to use black coal as fuel first appeared in China a few centuries before the English industrial revolution.

consumption growth till the stable point is much greater (see Fig. 5*a*). Thus, the two arcs together, joined in point 1, give the phase portrait of total energy use: M as a function of K, or $M = M_w + M_c$ as a function of $K = K_w + K_c$.

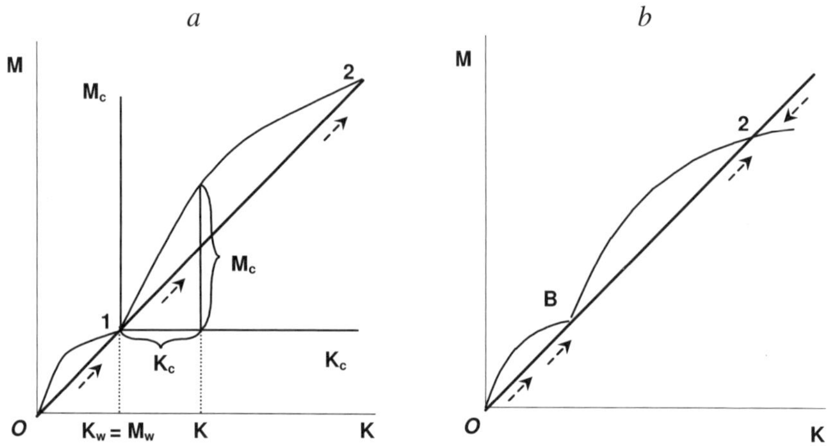

Fig. 5. Phase portrait of total energy use based on two power sources: fire wood and coal. Coal came into use after wood consumption had stabilized. Not to scale: the second arc should be many times greater.

It is more likely, however, that coal came into use to meet the growing fuel demand before the wood consumption stabilized. Then the sudden change in energy use would show up as a break point in the phase curve (point B in Fig. 5*b*) but the descending curve would not reach the bisector, i.e., energy use would rise continuously (Fig. 5*b*). Figure 6 including hydrocarbons (oil and gas) which have been in use for about 150 years already (third arc) and nuclear power (fourth arc) is a more realistic image of the today's total energy use than Fig. 5*b* showing only wood and coal fuels. The appearance of hydrocarbon fuels and nuclear power is well documented in statistics; it preceded the stabilization of each previous source in both cases.

The next arc may correspond to solar energy. Its conversion into electricity, too expensive yet, is the primary challenge of the modern technology. Note that biota solved this problem long ago by inventing photosynthesis to win the competition with chemosynthesis. Thermonuclear energy, with its inexhaustible resource, might be another arc and is a still greater challenge. There people will likewise follow nature: fusion reaction has been maintained successfully in stars, and in the Sun, for billions of years. Both sources are environment-friendly. The less important sources (hydro, wind, or tidal power) are not included in the figure.

Note again that unlike the case of Chapter 1, we are discussing the behavior of a special taxon, such as the humans. The use of all fuels taken together grows continuously and no value achieved in any current year ever repeats, so that we have a single "standard experiment" in terms of the modeling of Chapter 1. The prediction of M in the year following K would hardly be reliable even if the technological advance broke and energy use fell back to an earlier level as a result of some catastrophic event. Unlike animals, people never return to a situation they had lived through before as they ever change themselves, their ways and facilities. For instance, Germany and Japan, the most heavily ruined

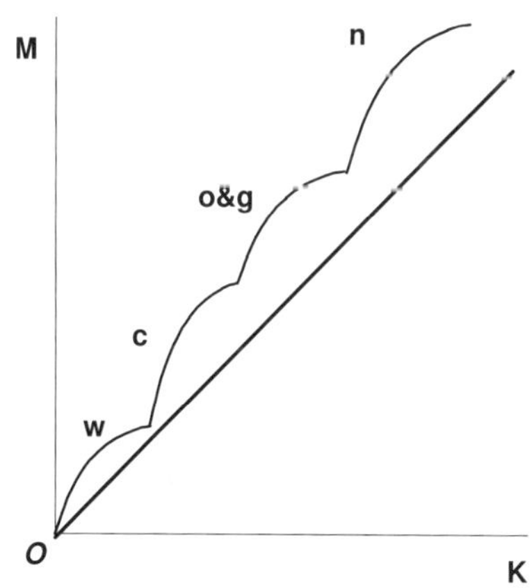

Fig. 6. Phase portrait of total energy use based on four power sources: fire wood (1), coke (2), hydrocarbons (3), and nuclear power (4).

after the second world war, returned to a lower level of energy production but the further evolution was more rapid. Therefore, phase portraits provide only a generalized qualitative image, even for global-scale processes, and are tentative especially for human activities. In the early 1900s coal resources seemed about exhaustion and the global energy use was expected to fall down (like the pattern in Fig. 5*b*) but coal was soon found in abundance and more efficient energy sources were discovered later (Fig. 6). They are rather environmental problems than energy deficit that threaten the progress of technology.

Nevertheless, the energy phase portraits of Figs. 5 and 6 show a certain tendency of energy use: the convexity of all arcs records periods of slower growth. The growth inevitably slows down because, like in the case of animal populations, energy resources as well as life conditions are limited. In-field fossil fuels are abundant but their development is becoming ever more costly, which stimulates energy saving technologies. On the other hand, environmental constraints limit the growth of industry, especially the energy-consuming heavy industry. Thus energy use may stabilize in the future (as well as the world population). It has been growing linearly and not exponentially since the 1970s (Fig. 2), contrary to the past century fears. Of course, stabilization may be impeded by rising energy use in developing countries.

Greenhouse effect: controls from technology

The Earth receives a half of the total solar radiation, about 50,000 kW per man and about as much is emitted back to space, while the whole world industry gives less than 1 kW per man. Thus, the technology appears quite inoffensive against the cosmic processes being unable to cause any serious direct overheating, but the jeopardy is elsewhere.

If atmospheric greenhouse gases increase for technological reasons and thus absorb and re-emit back to the surface more outgoing terrestrial infrared radiation for a value ΔW, the total received radiation increases correspondingly for ΔW and, by the conservation law, the terrestrial radiation increases for the same magnitude. The related planetary temperature rise (ΔT) is predictable from the Stefan–Boltzmann law (see Chapter 3). Thus temperature changes caused by the known changes in air composition can be predicted using the rigorous methods of physics.

The greenhouse effect is not a recent phenomenon due uniquely to technology and is beneficial by itself. Without the greenhouse gases, the Earth's surface would be about 33°C colder and unable to develop and support a biosphere. What is really worrisome is its overwhelming growth since 1750, or during the industrial era.

Carbon dioxide, though a less efficient absorber of the outgoing long-wavelength radiation than methane (CH_4) and nitrous oxide (N_2O), accounts for about 60% of the total greenhouse effect being a larger constituent of the modern atmosphere. The content of atmospheric CO_2 recorded in Antarctic and Greenland ice cores was more or less invariable from the Last Glacial (about ten thousand years ago) to 1750; since then it has been increasing exponentially, from 270 ppm to 350 ppm (part per million) in the 1990s (Fig. 7).

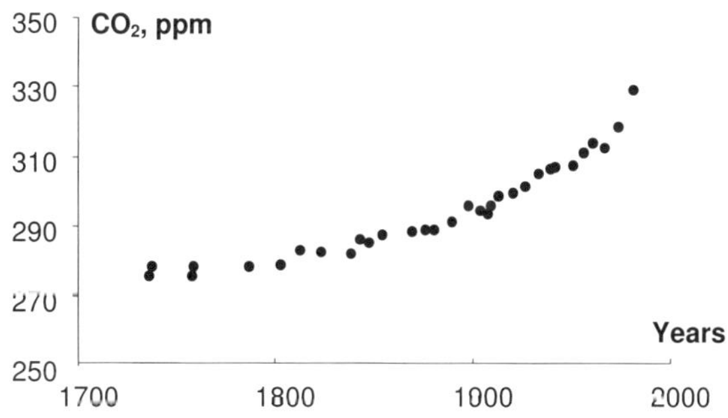

Fig. 7. Atmospheric CO_2 since 1750, from ice core data, Siple glacier, western Antarctica.

Exponential growth of atmospheric CO_2 is reported from many geographically dispersed observatories worldwide (Fig. 8) [*Semenov*, 2004]. The estimate of total annual increase of about 0.4% or ×1.004 causes no dispute. If the growth rate holds, the CO_2 content will reach 500 ppm by 2100, or twice the pre-industrial level.

The CO_2 trend correlates with the growth of fossil fuel supply which doubled every decade till recently. Industry annually releases into the atmosphere about 7×10^{12} kg carbon in CO_2. About a half becomes involved in seawater, photosynthesis, and other natural processes (e.g., peat deposition). This portion remains more or less invariable and will hardly increase in future. Therefore, if the tendency keeps up, the exponential growth of Fig. 8 will continue.

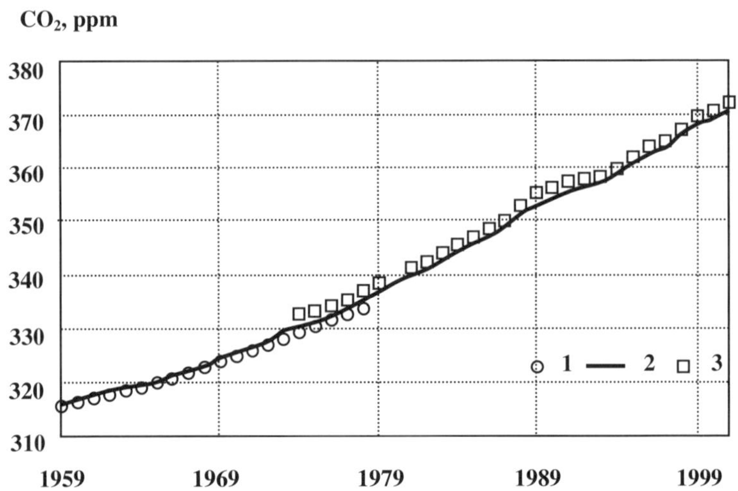

Fig. 8. Growth of atmospheric CO_2 from 1959 to 1999 measured at different monitoring stations: 1 = Law Dome, Antarctica; 2 = Mouna-Loa, Hawaii; 3 = Barrow, Alaska. Data from the Law Dome station are CO_2 concentrations in air bubbles in ice cores; data from the two other stations are mean annual CO_2 concentrations in near-surface air, from instrumental measurements (after *Etheridge et al.*, 1998; *Keeling and Worf*, 2004.) Borrowed from *Semenov, 2004*.

Exponential growth is rarely found in nature but is very usual in human activities. It corresponds to positive feedback when an unrestrained increase in some factor escalates its further increase. Positive feedback works in hazards such as forest fires or avalanches. The exponentially growing fossil fuel and atmospheric CO_2 are obviously related, and the former controls the latter.

Atmospheric methane, another greenhouse gas, comes from natural sources and is emitted from coal mines, natural gas wells, and pipelines. The methane level doubled during the industrial era from the early 19th century to 1980s and increases for 1% annually. A methane molecule uptakes about 70 times more long-wavelength radiation than CO_2 but less than an N_2O molecule (the multiple is 250 times) coming to the atmosphere from fertilizers and biomass, and fossil fuels combustion. Concentrations of atmospheric N_2O rose from 300 to about 310 ppt (parts per trillion) between 1977 and 1990 [Smil, 1994]. Besides the natural greenhouse gases, infrared radiation is absorbed by manmade chlorofluorocarbon compounds (CFCs) used since the 1930s as coolants in refrigerators, aerosol propellants, foaming agents, and cleansers. CFC molecules are the most efficient absorbers of the terrestrial radiation, 10,000 times more efficient than CO_2 (for more details of CFC use see Chapter 5).

Temperature rise and climate change

Speaking about Earth's surface temperature we mean the general globally averaged trend over long periods of time without casual excursions, i.e., the equilibrium temperature as is used in the Stefan–Boltzmann law (see Chapter 3). Then, it is much easier to predict, say, the average between 2040 and 2060 than the mean annual temperature of 2050.

Note that temperature estimates are intrinsically uncertain because of water vapor content variations. Temperature rise by greenhouse gases accelerates evaporation, and vapor, in turn, causes an additional greenhouse effect [Del Genio, 2002], which eludes exact assessment. Therefore, long-term temperature predictions as a rule give the lower and upper limits.

Reliable air temperature accounts are available since 1860 and show a mean global rise of 0.3°C to 0.6°C (~0.003 annually). This would appear quite small, but the total natural growth has been 5°C over the past 10,000 years (0.0005 annually), or about an order of magnitude slower. The greenhouse effect is greater on continents than in the oceans and thus in the more continental Northern hemisphere. Climate studies indicate an indubitable general warming trend through the past century showing up as a greater annual number of frost-free evenings, a longer vegetation period of plants, warmer winters in Siberia and elsewhere, etc.

The UNO Intergovernmental Panel on Climate Change that involves about 2000 scientists from many countries predicts an at least 1°C and up to 3.5°C rise in the mean annual global air temperature by 2100 unless the emission of greenhouse gases strongly reduces. The greatest warming is expected between 40° and 70° N where the rise was the highest in the 20th century.

Warming by itself is far less dangerous than its consequences which are more difficult to predict. The Earth is an extremely complicated system, and numerical modeling applied to climate studies is less reliable than the physical methods. To understand the difference imagine a person cooking soup: It is quite easy to know how much heat the pan receives but hard to predict whether the soup will be good. Nevertheless, there are credible predictions based on mathematical modeling and simulation. Numerical models are tested against measured data for past periods of time and are considered satisfactory and applicable to forecasting if the computed behavior of the input parameters fits the observations. Yet, the modeled processes are assumed to keep to the same course.

The temperature rise due to technological emission of greenhouse gases at the current rate may push forward thawing ice sheets and mountain glaciers. Thawing of the largest ice sheets in Antarctica and Greenland will cause a sealevel rise and the ensuing flooding of densely populated territories in northern Europe, Bangladesh, China, etc., now inhabited with a total of about 90 million people, and many ports, including Russian St. Petersburg. The sealevel rose for at least 10 cm over the past century and is now rising at 2 mm/yr; if this rate holds, the sealevel will be 50 cm higher in 2100. No less important consequence is that ice sheets waning decreases the total global albedo and increases the share of solar radiation absorbed by the Earth's surface correspondingly, which causes additional warming. Thus positive feedback comes into play, and its effects are hard to predict. The global albedo has already decreased by the disappearance of the Kilimanjaro glacier in Africa, where the incoming solar rays are normal to the surface and heat it up the most strongly.

The ways of further evolution can be discussed in the context of the atmosphere-biosphere interaction where technology became another regulator, primarily due to the manmade aggravation of the greenhouse effect. Now we summarize the model of the life-atmosphere interplay (for details see Chapter 3) with regard to technospheric forcing. The effect of the Earth's atmosphere with its greenhouse gases (water vapor, carbon dioxide, methane, etc.) on the surface temperature is plotted in the coordinates *C* (CO_2) and *T* (temperature), see curve 1 (*T*-nullcline) in Fig. 9*a*.

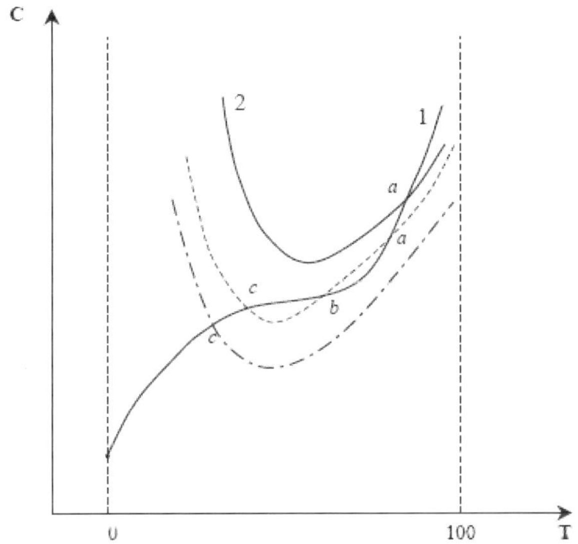

Fig. 9*a*. Effect of the Earth's atmosphere with its greenhouse gases on the mean global surface temperature plotted in the coordinates *T* (temperature) and *C* (CO_2). Life-atmosphere balance controlled by evolution.

The intersection of the *T* and *C* nullclines (point *a*) defines the stable state of the temperature and CO_2. As far as life evolves and the total volume of biota increases, curve 2 progressively moves down (dashed line in Fig. 9*a*) at a rate corresponding to the geological time scale. Later on the downgoing *C*-nullcline (curve 2) crosses the *T*-nullcline (curve 1) at two more points (*b* and *c*). The point *c* is an equilibrium state of temperature and CO_2 corresponding to a colder global climate and *b* is an unstable saddle or a divide between warm and cold climates. Further evolution of life can move curve 2 still lower (chain line in Fig. 9*a*).

Then the two curves again cross at a single equilibrium point (c) of a cold climate which can be either stable or unstable (see Chapter 3). The transition from a to c means the change from a warm climate (like that of, say, Tertiary time) to a colder climate (like that of nowadays). Transitions back to a warm climate were possible in the Earth's history when atmospheric CO_2 increased due to geological events (primarily, volcanic activity) which slowly moved curve 2 up. Repeated alternation from warm to cold climate and back through the Earth's history left authentic signature in paleontological and geological records.

The interference of man into the biota-climate balance which has equilibrated in the course of million-years long evolution acts as intervention of a rapid process into the interplay of two processes with many orders of magnitude greater characteristic times. The two processes of biota and temperature evolution have defined the basic phase portrait of the system, namely the number of its "anchor" equilibrium points. The relatively short-term technological climate forcing can push the life-atmosphere system off the equilibrium point(s) it currently occupies and make it move within its broad phase portrait.

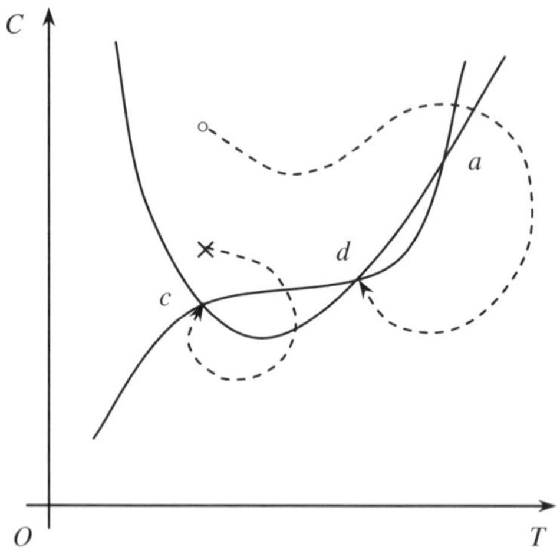

Fig. 9b. Interference of technology in the case of three equilibrium points of the life-atmosphere system. Cross marks the point where the system can move at a moderate CO_2 growth. Circle marks the point where the system can move at a rapid CO_2 growth.

This motion can follow several scenarios. (1) If the system occurs at the point *c* (corresponding to cold climate) and the two nullclines cross at three points, relatively moderate increase in atmospheric CO_2 will not cause dramatic global change as the climate will turn back to *c* in several hundreds of years (see the cross in Fig. 9*b*). (2) However, exponential CO_2 growth can displace the system as far from the point *c* as to make it leap over the divide *b* (see the circle in Fig. 9*b*). Then, the system won't be able to return to *c*, even if CO_2 release stops, and the evolution will bring it to the stable point *a* of a warm climate like that of Tertiary time when a great part of land was swampy and the sea stand was much higher. (3) If the two nullclines cross at a single point (*c*), the system pushed off its equilibrium will inevitably return to *c*. Yet the recovery time and the amplitude of temperature change depends on the amount of carbon dioxide. A too rapid CO_2 growth can move the system very far and make the amplitude of climate oscillations high enough for biota and sealevel changes to be dramatic (see the cross and the circle Fig. 9*c*). The

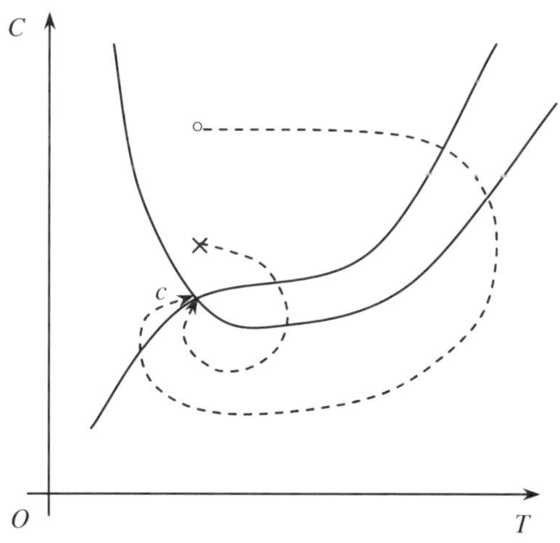

Fig. 9*c*. Interference of technology in the case of a single equilibrium point of the life-atmosphere system. Cross marks the point where the system can move at a moderate CO_2 growth. Circle marks the point where the system can move at a rapid CO_2 growth.

motion of the system along the paths off the equilibrium point *c* is very rapid on the geological scale of time but slow on the scale of a human life. The related climate change can occur within a century and turning back to the equilibrium can take thousands of years, the time CO_2 and other technologically released greenhouse gases can persist in the atmosphere. Therefore, many generations of people will live in a different climate anyway.

This warning stems from the qualitative theory of differential equations the prediction of which have an unpleasant property of being inescapable.

Unfortunately, nobody knows whether there are one or three equilibrium points of the biota-climate system at present but it is possible — and urgent — to estimate the critical total amount of greenhouse gases that can trigger the leap over the divide point, if there is one.

One should be aware that CO_2 pollution may be irreversible. No technologically practicable ways are currently available to remove the billions of tons of carbon dioxide let into air for over 250 years. Binding this gas would require an amount of energy comparable to that released in its production. Cleaning the Earth's atmosphere from manmade carbon dioxide is a challenge like building a Marsian atmosphere or similar projects feasible so far only in science fiction.

The greenhouse effect is quite different from any other consequences of man's activity as there people switch a cosmic process. The question is whether people want the changes they call forth and whether they take on this responsibility before the coming generations.

Alternative power sources

Preventing our home planet from surpassing the divide to a warm climate is a common concern. Plausible solutions to the problem may require some limitations in the use of energy resources and making the alternative power sources really competitive to the fossil fuels.

Nowadays nuclear power is the only serious competitor of fossil fuels. The prejudice of public against nuclear power is reasonable being due to the tragedies of testing the nuclear bomb in Japan and the Chernobyl accident in the Soviet Ukraine caused by the criminal negligence of officials. This prejudice is so vigorous that people miss important facts.

Leaving aside the financial details, the use of nuclear power, including its transportation, costs roughly as much as fossil fuel use and is advantageous for many countries. For example, nuclear power stations in France generate 75% of electricity. With the appropriate precautions, nuclear power stations are safe even in their immediate vicinity. Their radiation is below the natural background and below the doses received in a medical X-ray examination. A power station 60 km upstream of Lyons on the Rhone has caused no troubles for decades. Outside the former Soviet Union there were no fatal accidents over forty years of nuclear power production.

Nobody cares too much about billions of tons of much more radioactive wastes from fossil fuels because they mostly dissipate in air remaining invisible. Unlike the fossil fuel wastes from thermal stations, the less voluminous nuclear wastes are collected in tight containers and are made isolated. Naive people cannot imagine that a threatening "thousand tons nuclear wastes" chewed over in mass media makes just a 10×10×10 m cube, and almost all wastes can be recycled and reused. Storage of nuclear wastes buried using advanced technologies can cause no problem for hundreds of years, and the work is underway to ensure safe storage for 10,000 years. It sounds reassuring but vigilance and public control are required. In a sense, it is another problem we pass on to the coming generations, not as an irreversible contamination of the atmosphere but as small guarded depots in dry salt mines or on remote islands.

Nuclear power generation has, however, some drawbacks. First, it is the problem of putting stations out of operation. Some have been in operation for over forty years and the takedown may cost more than their building.

The safety of nuclear fuel is another problem. Uranium mines require special caution measures as natural U ore itself is highly radioactive. The appropriate safety measures for people engaged in mining and transportation make nuclear fuel more costly and less payable. On the other hand, insufficient measures make mining fatal for people, especially in the countries that lack law and order. Unfortunately, reliable information on uranium mining is unavailable to public.

So far all efforts and expenses have addressed the fossil-fuel and nuclear power supply which as yet cannot give way to environment-friendly power engineering using alternative sources.

Wood is an environmentally safe power source, provided that forest is recovered by plantations and special protection measures, because growing trees bind as much carbon dioxide as is released in burning. Yet this source is far insufficient for the modern technology. Unlike wood, coal and oil are unrecoverable resources.

Hydro power stations which are profitable only if set up on rapid highland rivers or wind stations which depend on weather caprices are workable rather for local uses. The future of power engineering lies with solar and, perhaps, thermonuclear energy if physicists cope with the latter.

Solar arrays are advantageous as they lack dangerous components, are safe in fabrication and release nothing to the atmosphere. So far the use of solar energy has been restricted to local household-scale applications in some southern countries but one can expect that a part of a desert covered with mirror-like solar arrays will in the future safely meet the demand of a large economy. The future power engineering needs new principles as the modern technology, whichever be its accomplishments, is at a deadlock. The first priorities are transportation and storage of energy and nuclear fusion.

The existing transportation ways are inefficient because of large losses from power lines: transport for over 2000 km is uneconomic. Carrying coke and crude oil by railway looks still more archaic. The design of the available electric batteries is just a modification of old ideas. No portable and safe engines have been invented; the old combustion engine is likewise being improved in details but remains basically the same as an age ago.

Managing thermonuclear power sources which have been operating successfully in stars for billions of years is especially challenging. Physicists had never faced a problem as hard as controlled thermonuclear fusion (CTF). The main problem is that CTF requires high-temperature plasma to be restrained from spreading and, at the same time, insulated from the reactor walls. Sustaining the reaction in the terrestrial conditions requires heating plasma to as hot as 100–200 million degrees. The numerous systems for confinement of high-temperature plasma either employ inertial confinement using compression by X rays, ion or laser beams, or thermal insulation by magnetic field. Yet, a hydrogen bomb has been the only practical application found for the inexhaustible and environment-friendly energy source of nuclear fusion. Note for comparison how much time elapsed between the theoretical discovery of nuclear energy (by Einstein, 1905) and Fermi's reactor (1942), likewise applied to make a bomb. People have enough bombs and would rather welcome more useful accomplishments of physicists.

There are no valid reasons to risk the global atmospheric balance by pushing forward fossil-fuel power engineering. Perhaps, the use of fossil fuels was historically necessary but continuing the practice instead of looking for its alternatives is an inexcusable error. Now people have to recover the major faults and especially avoid new ones.

At all stages of the energy use in the world history, new sources were sought and found only after the previous sources became exhausted or the need was imposed by economy. The current situation is unique in the urge to put alternative power sources into use though fossil fuels are still in abundance and are especially attractive being highly economic. Thus the primary task of international organizations like UNO and individual states is to eliminate the economic attractiveness of fossil fuels, for example, by strongly increasing payment for their use. Otherwise, the use of fossil fuels dooms people to an environment collapse associated with global warming and its consequences. On the other hand, refusing the advance of technology as suggested by some green-minded people means refusing a working economy which is fatal for the numerous population of the planet. The way we suggest does not mean locking the progress but requires a control of reason over the advance of technology.

REFERENCES

Del Genio A. D., 2002. The dust settles on water vapor feedback, *Science*, 296(5568), 665–666.

Etheridge D.M., Steele L.P., Langenfelds R.L., Francey R.J., Barnola J.-M., and Morgan V.I., 1995. Historical CO_2 records since about 1000 a.d. from ice core data, in: Trends: A compendium of data on global change, Carbon dioxide information analysis center, Oak Ridge National Laboratory, US Department of energy.

International energy outlook, US Department of Energy.

Keeling C.D. and Whorf T.P., 2004. Atmospheric CO_2 records from sites in the SIO air sampling network, in: Trends: A compendium of data on global change, Carbon dioxide information analysis center, Oak Ridge National Laboratory, US Department of energy.

Semenov S.M., 2004. *Greenhouse gases and present climate of the Earth*, Meteorologiya i Gidrologiya, Moscow, 175 pp. (in Russian).

Smil V., 1994. *Energy in world history*, Westview press, Boulder-San Francisko-Oxford, 300 pp.

UNO Energy statistics yearbook, 1982; 1992.

Chapter 5

Dynamics of Atmospheric Ozone

Life endowed the Earth's atmosphere with free oxygen and free oxygen became the source of ozone. Atmospheric ozone cuts off the short wavelength portion of the solar spectrum, or the ultraviolet (UV) radiation, which destroys proteins and nucleic acids thus making the land life impossible, at least in its present form. Prior to the advent of photosynthesis and the ensuing appearance of atmospheric free oxygen, life could exist only in the water as the hard UV radiation reached the Earth's surface and killed all living that dared to emerge. Having created the ozone layer, life allowed itself to expand over the land. Therefore, the time when plants settled the land habitats corresponded to the origin of the ozone layer.

The chemically unstable ozone molecules naturally form and break down under the effect of biotic and abiotic agents. The evolution has brought the process to some balance but technology has been interfering with this balance. The consequences, showing up as depletion of the ozone layer and reducing its screening capacity, were revealed in the second half of the 20th century and alarmed scientists and then the broad public. The question has been whether and how much technology is responsible for the ozone depletion and the appearance of ozone holes.

In this chapter we address the environmental problem of atmospheric ozone and mechanisms of its dynamics. We investigate the formation of large ozone holes over the south pole and smaller holes in the middle latitudes of the northern hemisphere using a new method suggested by V.B. Kashkin for tracing stratospheric air flows.

Ozone shield

Ozone is one of the several gases that constitute the Earth's atmosphere. It makes up approximately one part in three million of all of the atmospheric gases. If all the ozone contained in the atmosphere from the ground level up to a height of 60 km could be assembled at the earth's surface, it would comprise a layer of gas only about 3 millimetres thick. But even though ozone occurs in such small quantities, it plays an exceptional part in life on Earth protecting it from the hard ultraviolet radiation fatal for living organisms. Ozone, together with ordinary molecular oxygen (O_2), is able to absorb the major part of the incoming solar ultraviolet light and thus prevent it from reaching the surface. Ozone absorbs about 5×10^{20} J/day or 3% of the total solar energy flux, which is equivalent to the energy of about one thousand tropospheric (lower atmospheric, below 10 km) cyclones.

Stratospheric ozone is an important climate agent responsible for short-term and local weather variations. Absorption of solar radiation by ozone and energy transfer to other gases causes quite strong heating of stratosphere and thus controls planetary-scale thermal and circulation processes in the atmosphere.

Ozone is a specific triatomic form of oxygen: an ozone molecule O_3 consists of three oxygen atoms. Almost all atmospheric oxygen (20.95%) exists as stable molecules of O_2 while O_3 is unstable. Ozone molecules live for about ninety days and then break down into atoms which then reassociate into ordinary O_2.

Ozone was identified as a separate chemical substance in 1840 by Christian Friedrich Shoenbein, a Swiss-German chemist. In 1873 ozone was discovered in the lower atmosphere, and in 1881 Walter Hartley, an Irish chemist, predicted its presence in the upper atmosphere from absorption of solar radiation. Pure ozone was obtained in 1922 by E.H. Riesenfeld and G.M. Schwab.

The English physicist Sidney Chapman formulated in 1930 the first photochemical theory of the formation and decomposition of atmospheric ozone. This theory describes how sunlight converts the various forms of oxygen from one to another: ultraviolet radiation from the sun (at wavelengths below 0.242 μm) splits molecular oxygen and the oxygen atoms thereby liberated react according to $O_2 + O + M \rightarrow O_3 + M$, where M is any random air molecule (N_2 or O_2) that takes on excess energy. The reaction can occur at heights of 25 to 70 km because the UV radiation required for dissociation of molecular oxygen does not reach lower heights.

Atmospheric ozone exists in an about 90 km thick spherical layer and its inner surface coincides with the Earth's surface. Nearly 90% ozone occurs in the stratosphere at a height of 10 to 50 km. The average density of ozone in the layer from 0 to 70 km is 0.9×10^{-10} g/cm^3. It is distributed unevenly, both vertically and laterally. The density of the atmosphere and the content of oxygen required for the formation of ozone decrease with height while the intensity of UV radiation increases. Therefore, the ozone concentration is maximum at a height of 26–27 km in tropic latitudes, at 20–21 km in middle latitudes, and at 15–17 km in the polar regions.

Ozone is a variable air component. In the average it makes only $4\cdot10^{-5}$ vol.% and its total weight is about 3×10^{12} kg, or 0.64×10^{-6} total atmosphere weight. Total column ozone (TCO), or total amount of ozone above the station, is subject to diurnal, seasonal, annual, and decadal variations. Mean global TCO is about 290 Dobson Units (DU). 100 DU is equivalent to a 1 millimeter thick layer of pure ozone at the sea level temperature and pressure, and TCO thus actually measures the ozone layer thickness.

The observed natural thickness of the ozone layer varies in a broad range from 90 to 760 DU, and deviation from the global mean can reach 25% within a day. Annual total ozone variations are the least near the equator (within 250–280 DU). The highest total ozone (500–700 DU) in the middle latitudes of the northern hemisphere is recorded in spring and the lowest in autumn.

Total ozone (TCO) is measured using the optical properties of ozone that absorbs and backscatters solar radiation. The instrument, a Dobson spectrophotometer, compares the amount of sunlight at two (or four) ultraviolet wavelengths. This allows excluding the effect of aerosols and other atmospheric disturbances and factors associated with variations of radiation at the upper atmosphere boundary. The world network included 131 land ozone monitoring stations in the 1980s, scattered very unevenly over the continental area. This network has very little chance to detect anomalies of total ozone even if the latter have a linear dimension of 1000 km.

A higher-density monitoring is provided by optical instruments mounted on satellites. Satellite UV spectrophotometers measure radiation backscattered by ozone molecules. TOMS spectrophotometers (USA) use four wavelengths. These instruments were successfully operated with the Russian satellite Meteor-3, with the Japanese ADEOS, and with the American TOMS/EP; at present a special ozonometric satellite AURA is under the US operation.

Ozone layer destruction

Ozone, as well as molecular oxygen, is subject to external effects (radiation or particle fluxes) which cause its decomposition. When absorbing UV or visible light at an energy quantum over 1.09 eV, an ozone molecule splits into an O atom and diatomic oxygen. It was found out that ozone destruction is strongly controlled by the presence of substances that react catalytically (without themselves being consumed) with ozone, thus accelerating the rate of reduction of the ozone content. The catalysts can be either natural oxides inherent to the atmosphere or those added as a result of natural hazards like volcanic eruptions, or by technological activity. Ozone destruction is caused by oxygen, nitrogen, hydrogen, or chlorine photochemical cycles. The highly reactive oxygen atoms and ozone molecules can interact as $O + O_3 \rightarrow O_2 + O_2$ (oxygen cycle). The nitrogen cycle reactions are $O_3 + NO \rightarrow NO_2 + O_2$, $NO_2 + O \rightarrow NO + O_2$, in which O and O_3 loss can be times faster when the concentration of NO and NO_2 is high enough, especially in the layer of maximum ozone content.

Hydrogen and its oxides (hydrogen cycle) is an especially active agent for lower stratospheric ozone and atomic oxygen: $H + O_3 \rightarrow OH + O_2$, $O + OH \rightarrow H + O_2$ $OH + O_3 \rightarrow HO_2 + O_2$, $HO_2 + O_3 \rightarrow H + 2O_2$.

Stratospheric compounds of Cl can participate in over seventy reactions of ozone dissociation (chlorine cycle) where ClO acts as a catalyst (Br behaves in the same way): $Cl + O_3 \rightarrow ClO + O_2$, $ClO + O \rightarrow Cl + O_2$. Cl atoms are released to the stratosphere in decomposition of chlorine compounds under the effect of sunlight.

Different estimates more or less agree about the percentages of ozone losses in the four cycles: 17% in the oxygen cycle, 9 to 11% in the hydrogen cycle, 70% in the nitrogen cycle, and 4% in the chlorine cycle; about 1.2% ozone goes to troposphere. The resulting loss is not the simple sum of the effects from separate cycles as these change by reactions between the members of different families.

In the 1970-1980s, some scientists concluded that chlorine and its compounds, which increase and accumulate in the atmosphere mostly as a result of industrial emissions, can be responsible for ozone depletion [Roan, 1989]. In 1973, American chemists F. Sherwood Rowland and Mario Molina discovered that Cl atoms released from manmade synthetic chemicals (especially chlorofluorocarbons or CFCs, also called freons) as a result of their breakdown under sunlight can destroy the stratospheric ozone layer. Rowland and Molina published their findings in Nature on June 28, 1974, and in 1995 they, together with Paul Crutzen, won the Nobel Prize in chemistry for their discovery.

The ideas were supported by the US National Academy of Sciences and received much attention. In 1975 the use of CFSs in aerosol sprays was banned on the initiative of the US National Resources Defense Council (NRDC) and they were eliminated in the United States by 1978. In 1976, 1979, 1982 and 1984 the US Academy submitted reports where predicted dangerous depletion of the ozone layer by CFCs. First international efforts for ozone protection were undertaken in 1977 in Washington when representatives of 32 countries prepared a phase-out treaty for restrictions on CFCs and other ozone-depleting chemicals. On March 22, 1985, twenty nations signed the Convention for the Protection of the Ozone Layer in Vienna. In 1986, NRDC proposed a global 10-year phase-out plan to replace CFCs with alternatives safer for the ozone layer. A similar approach formed the basis of the Montreal Protocol on Substances that Deplete the Ozone Layer, the first-ever global environmental agreement negotiated under the auspices of the United Nations and signed in 1987 by 57 industrial nations. Amended several times and now signed by more than 180 nations, the Montreal Protocol halted all production of CFCs in developed countries by 1996. Developing nations are now following suit, with a deadline of 2006. The Montreal protocol implies control over 95 ozone-depleting chemicals.

The environment protection law of the Russian Federation includes a special article for protection of the ozone layer. Production of synthetic ozone-depleting chemicals in Russia is progressively decreasing. It culminated in 1990 with almost 200 thousand tons, including 110 thousand tons CFCs and reduced in 1996 to 47 and 17 thousand tons, respectively.

When first synthesized, CFCs showed a number of advantages being non-toxic, highly stable, fire-safe, and compatible with many materials. They replaced earlier used toxic coolants in household refrigerators and found broad application in home and car air conditioners, as propellants for dispensing aerosol sprays, blowing or bubbling agents for foam packaging and insulation, and cleaning agents and in the manufacture of computers. The widest known freons are fully halogenated CFC-11 and CFC-12 which cause the greatest damage to the ozone layer.

Less offensive are hydrochlorofluorocarbons (HCFCs), which appeared long before people became aware of the ozone emergency, and the quite recently synthesized hydrofluorocarbons (HFCs). Although containing chlorine, HCFCs cannot destroy the ozone layer as their hydrogen atoms facilitate rapid binding of chlorine in the lower atmosphere; HFCs cause no damage to ozone being free from chlorine.

Prohibiting the manufacturing and use of ozone-depleting chemicals had serious political and economic consequences. The ban was first met with hostility from chemical industry people, especially in the US. Later Du Pont supported the restrictions for CFC production worldwide and developed HFCs as a CFC alternative. Then there followed a boom of replacing the old refrigerators and air conditioners by new ones free from ozone-damaging coolants. The companies that were the first to use new coolants got enormous gain. Moreover, the ban outlawing freons ensured monopoly of refrigerant production by the major chemical companies of the world (Du Pont, ICI, Allied Signal Inc. in the US, Hoechst AG in Germany, Atochem SA in France, and Showa Denko KK in Japan) which took hold of both patented products and their manufacturing facilities. Only the strongest could survive the competition. Losses from the CFC ban amounted to many tens of billion dollars only in the US. The Russian economy likewise incurred some damage though not so large.

There appeared ideas (see, for instance, "The Holes in the Ozone Scare: The Scientific Evidence That the Sky Isn't Falling" by Rogelio Maduro and Ralf Schauerhammer, written and published in 1992) that the ozone-saving policy might have been inspired by the greatest chemical companies like Du Pont and ICI to fight their competitors and to gain more strength in the world market. Maduro and Schauerhammer claim that the Montreal Protocol lets a minor technocratic elite dictate the economic policy to all nations.

Note anyway that the HCFC and HFC coolants are inferior to CFCs in many aspects. HFC-134a suggested by Du Pont is three to five times more expensive that CFC-12. Unlike CFCs, HFCFs and HFCs can become inflammable when mixed with air at high pressures and temperatures; furthermore, they are toxic and can cause corrosion. The use of HCFC and HFC instead of CFS coolants decreases the efficiency and durability of refrigerators and air conditioners and increases their price and power requirement, and also causes technological difficulties in manufacturing.

The problem whether Du Pont and other companies gained from the CFC ban is simple: of course they did. But this is a minor concern for the mankind. What really matters is how real has been and will be the emergency from the CFC use, how far could people mitigate the danger — if a danger did occur, — and whether the chosen measures have been successful or the alternatives worsened the environmental and economic standards of life.

Ozone holes

The problem of atmospheric ozone has not been resolved by the moment, and its many important issues are open to further research. Note that a slight ozone layer depletion is not a disaster, especially in the middle and high latitudes. Besides ozone, UV radiation is absorbed by clouds and aerosols. A large part of middle latitudes of the northern hemisphere, for example Central Siberia, where the number of cloudy days is up to 80% a year, suffers an ultraviolet deficit (about 45% of the medically required dose). On the other hand, total ozone in the equator, where Sun rays are normal to the atmosphere and travel the shortest way to the surface, is lower than in the middle and high latitudes (except for the Antarctic ozone hole) where the lower-angle sun rays are less intense.

The Antarctic ozone hole — the discovery of which stirred up public opinion on the ozone problem — usually appears every two years in the average, holds for 90–100 days and then disappears. The "ozone hole" is defined as the area with a substantial reduction below the naturally occurring concentration of ozone in the overhead column over a large territory. It is not an "open hole" but rather a sag of the geometrical surface drawn as DU ozone as a function of geographic coordinates. The sag forms in September as a low-total ozone zone surrounded by a

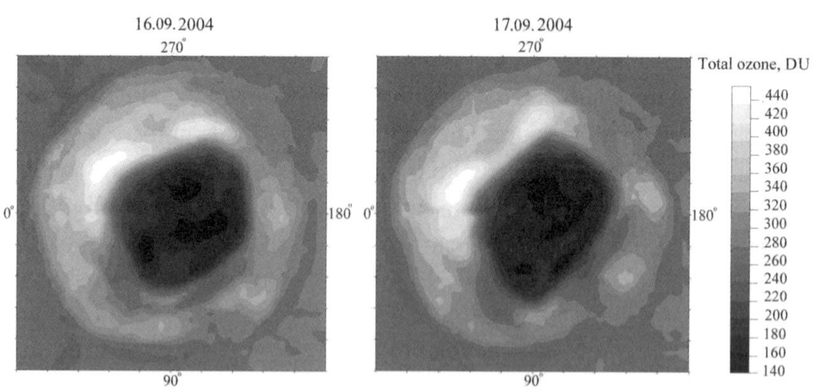

Fig. 1. Antarctic ozone hole of 16 and 17 September 2004.

high-ozone circumpolar vortex. See TOMS/EP satellite images of the ozone layer in the southern hemisphere on 16 and 17 September 2004 in Fig. 1. The total ozone field looks like a high-ozone ring (up to 460 DU) about 8000 km in diameter that circles a low-ozone area (the hole).

The notable depletion of the ozone layer was first discovered in 1957 during the International Geophysical Year. The same effect may have occurred before 1957, or perhaps has been a usual event. The true history of the Antarctic ozone hole began with the paper by Joseph Farman, Brian Gardiner, and Jonathan Shanklin, British scientists of the British Antarctic Survey, published in *Nature* [*Farman et al.*, 1985] which reported ozone data from two Antarctic stations since 1957. The stations recorded a drop in total ozone values since 1982 over Antarctica in southern hemisphere spring. Farman et al. were the first to hypothesize that industrially manufactured gases, including CFCs, may be responsible for the spring ozone loss over Antarctica. Further studies confirmed ozone depletion over Antarctica in southern hemisphere late winter and spring. For instance, an exceptional low of 88 DU was recorded on 28 September 1994 near the south pole and the ozone hole reached 30×10^6 km^2 in area. The 2000 ozone hole was as large as that and 20% larger than the 2004 hole. In 2003 there was almost no ozone hole. Unfortunately, the studies of the Antarctic ozone hole mostly aimed at proving its man-induced origin [*Roan*, 1989].

Formation mechanisms of ozone holes: hypotheses

The existing hypotheses on the formation mechanism of ozone holes attribute its origin either to manmade or natural causes.

One manmade-cause scenario for the Antarctic ozone hole invokes an important part of the lower stratospheric circumpolar vortex [*Farman et al.*, 1985]. It is assumed proceeding from theoretical ideas that the vortex breaks the ozone supply to the polar region while industrially manufactured chemical agents destroy the ozone that existed inside the vortex.

The theory of the manmade chemical origin of the ozone hole with participation of CFCs is not universally supported as it faces a number of contradictions. The most naive question is why should the hole exist in the southern hemisphere when CFCs are produced in the northern one, more so that there are no winds from the northern to southern hemisphere. Furthermore, the 1979-1995 period of strongest ozone depletion was punctuated by higher-ozone spells, e.g., in 1988 after a very deep ozone hole a year before. Moreover, the CFC chemical theory fails to explain the locally observed increase in stratospheric ozone as the atmospheric CFC contents should build up and only destroy the ozone layer.

Another hypothesis takes into account natural chemical mechanisms. There is experimental evidence from the 1986 NASA ER-2 Aircraft stratospheric studies that ClO inside the ozone hole is much higher than on the periphery [*Roan*, 1989]. Yet, ClO not necessarily comes from CFC decomposition in the northern hemisphere. It may likewise come, together with other gases, from eruptions of Erebus active volcano located just in the region of the ozone hole [*Dibble*, 1989].

The chlorine cycle reactions that attack the ozone layer can be catalyzed by nitrogen compounds which likewise can originate by natural mechanisms. The polar stratosphere of the southern hemisphere develops clouds in winter (June–July) at –70°C or colder at heights of 15 to 22 km. The clouds descend to 10–18 km in August-September and disappear after the stratosphere warms up. These clouds consisting of ice crystals and drops of supercooled liquid may accumulate active nitrogen compounds and thus speed up the ozone-depleting chlorine cycle. Crystals of nitrous ice become larger and heavier as they accumulate more nitric molecules and sink deeper to the troposphere thus providing outflow of active nitric compounds from the ozone hole which increases the part of the chlorine cycle. The clouds dissipate and nitrogen compounds liberated on melting and evaporation of ice crystals again come into play in spring when the Sun rises higher and warms up the Antarctic stratosphere. Moreover, sunlight sets up photochemical ozone formation, ozone comes to the hole from the surrounding higher-ozone region, and the ozone hole heals up.

An alternative dynamic theory explains the origin of the Antarctic ozone hole in terms of stratospheric waves. The idea is that long-period temperature variations on the ocean surface induce atmospheric waves which can produce isolated Arctic and Antarctic stratospheric vortices and cool the lower stratosphere thus facilitating chemical ozone depletion. The waves are global-scale (thousands of kilometers) cyclones and anticyclones that strongly influence the stratospheric dynamics in the winter season. The atmospheric planetary waves are caused by time-dependent temperature contrasts produced by large mountain systems (Tibet and the Rocky mountains in the northern hemisphere and the Andes in the southern one) and the land/ocean interface. The temperature variations can change the wave activity, air circulation, and further isolate the winter-spring stratospheric polar vortices. Attributing the origin of the ozone hole to natural processes, the wave theory does not rule out the chemical mechanism of ozone depletion.

A new method for tracing stratospheric air flows (Kashkin method)

There is a question whether the Antarctic polar stratosphere is indeed isolated during the formation of the ozone hole. This problem, as well as many other related problems can be solved using the new method suggested by V.B. Kashkin for tracing stratospheric air dynamics [*Kashkin et al.*, 2002]. The density of stratospheric ozone existing as a large continuous cloud over the earth's surface varies from place to place and from time to time. This variation pattern is similar to density variations of water clouds on an overcast day. People have used the pattern of an overcast sky as indicator of tropospheric air flows (translation and rotation) since long ago.

The new method implies total measurements of ozone density in the stratosphere, for example, over the southern hemisphere, on a square grid (from place to place) at two successive moments of time (say, every 24 hours). Comparing the ozone fields of two successive days allows estimating the daily rotation angle, direction, and speed of ozone clouds.

These very parameters record the dynamics of stratospheric air. Thus the observed changes in the ozone layer pattern are used to recover translational and rotational motions of air in the stratosphere, i.e., to trace stratospheric winds and vortices, where ozone is used as a tracer.

The ozone field is divided into 4-5° rings centered on the south pole and the total ozone field in a selected ring on one day is correlated to the total ozone field rotated to the angle φ and shifted to the angle Δ toward or off the equator on the previous day in the same ring. Then all total ozone values within the ring are averaged. By plotting the relationship between the correlation coefficient R and the angle φ, one can find φ and Δ that correspond to the maximum R. The average angular speed of the W — E air transport (rotation) is $\omega = \varphi$ (deg/day), and the average speed of the N — S transport is $\upsilon = \Delta$ (deg/day). The greatest sample R occasionally reaching $R = 0.95$-0.98 fits the daily average of ω and υ.

The TOMS/EP images of Fig. 1 show that the circumpolar vortex is rotating and the ozone cloud inside the vortex is moving as well. In a day the ring between 45° and 50° S rotated westward to some angle and slightly deformed (see Fig. 1). Estimates by this method show that the ring turned through an angle of 13° and shifted 5° in the average toward the southern pole off the equator.

Thus, the new method provides a relatively simple and cheap tool for studying stratospheric processes and the state of the ozone layer.

Global stratospheric dynamics: facts

The new method was applied to investigate the ozone layer dynamics and related stratospheric air flows in 2000 when the Antarctic ozone hole was especially large. We [*Kashkin et al.*, 2002] used TOMS/EP ozone density measurements over the whole southern hemisphere from 0° (equator) to 80° S between 1 and 20 September and from 0° to 88° S in October. At that time the polar region was illuminated and TOMS/EP could measure total ozone. Note that the satellite data from the circumpolar region are of higher density due to the curvature of the globe surface.

Figure 2 shows the average speed of W — E (ω) and N — S (υ) transport estimated by comparing total ozone fields for every two next days between 1 and 20 September and the mean latitudinal total ozone distribution over the same period.

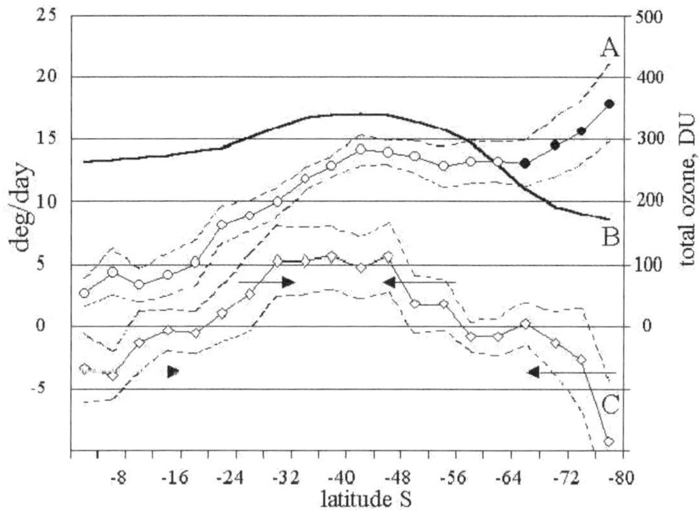

Fig. 2. Average angular speed ω of W — E transport (A) and speed υ of N — S transport (C) as a function of latitude (95% confidence limits), and mean latitude distribution of total ozone (B) between 1 and 20 September when the ozone hole was forming (DU scale on the right). Arrows show direction of N — S transport.

The N — S transport speed υ (curve C) is doubled for better illustration. Positive υ means that ozone arrives at some area and negative υ means that it leaves some area. See that ozone arrives at middle latitudes, which is the zone of the circumpolar vortex, and reaches the maximum total ozone (curve B).

Thus, in the presence of the ozone hole, ozone migrates to the middle latitudes from the equator as well as from the high latitudes.

The angular speed ω is time variable. See its variations in Fig. 3 between 1 and 20 September 2000 at 72° S, i.e., within the ozone hole (lowest ozone density, curve A) and at 48° S, i.e., within the circumpolar vortex (highest ozone density, curve B). Note irregular and quasi-periodic rotation of air at 72° S. The irregularity is of no surprise, more so that the rotation center is unstable and not necessarily coincides with the pole.

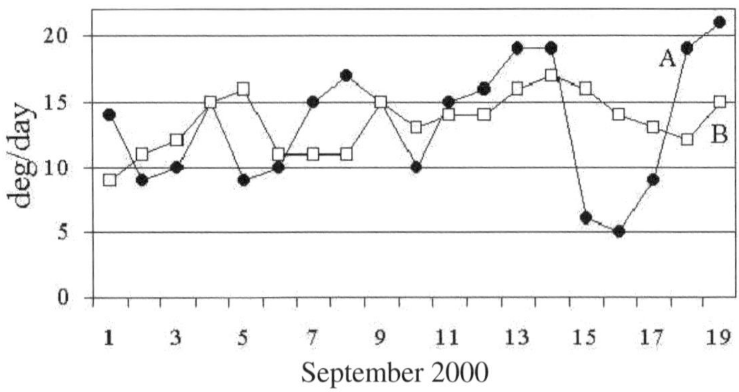

Fig. 3. Angular speed of ozone migration between 1 and 20 September 2000 at 72° S (A) and 48° S (B).

Of special interest is the latitude dependence of the angular speed ω. It increases away from the equator toward the middle latitudes and grows ever more as approaching the pole. Near the pole, between ~67° and ~80° S, its latitude dependence fits a parabola (dark circles in curve A, Fig. 2).

The parabolic dependence can be proven as follows. Consider two equal volumes of ozone of the mass m (1 and 2) inside a ring (Fig. 4). Their motion can be assumed independent of ambient air because of low atmospheric density at the height of the ozone maximum. Volume *1* is at the distance R_1 from the rotation center and the rotation speed is V_1; volume *2* is at the distance R_2 from the rotation center and the rotation speed is V_2. Assuming that the atmosphere behaves as a viscous liquid and taking into account the law of conservation of angular momentum, we can write: $R_1 \cdot m \cdot V_1 = R_2 \cdot m \cdot V_2$, but $V_1 = \omega_1 \cdot R_1$, $V_2 = \omega_2 \cdot R_2$, wherefrom $\omega_1/\omega_2 = (R_2/R_1)^2$, i.e., the dependence of the angular speed on the distance from the rotation center does fit a parabola, as in Fig. 2. Furthermore, this dependence keeps its parabolic shape outside the ozone hole between 10° and 40° S. Note that the parabolic dependence is typical of water whirlpools.

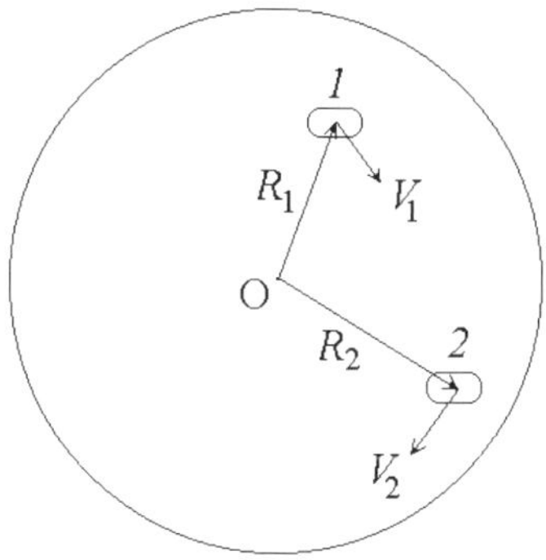

Fig. 4. Two equal volumes of ozone (*1* and *2*) rotating at V_1 and V_2 about the point O.

The ozone hole was the deepest on 29-30 September 2000 when the photochemical ozone formation already occurred inside the circumpolar vortex; after 30 September ozone showed persistent polarward migration. On some days ozone moved back off the pole, but eventually the hole healed up and disappeared. On 21–22 October, the hole became shallower and the maximum of total ozone shifted to higher latitudes. Compare Fig. 2 with Fig. 5 which shows the latitude dependences of the

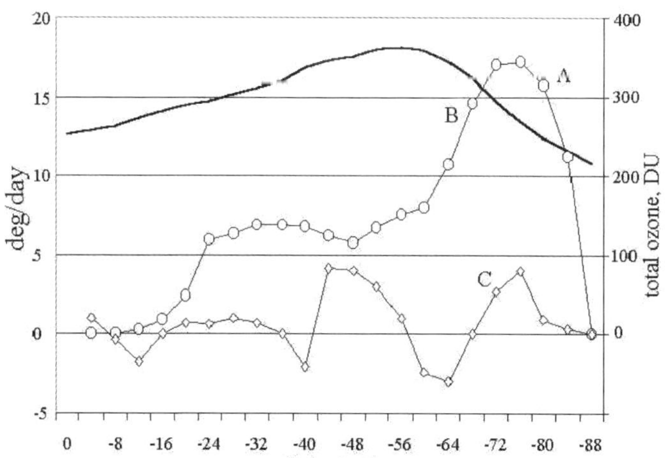

Fig. 5. Latitude dependences of the angular speed of W — E transport (curve A), total ozone (curve B), and speed of N — S transport (curve C) averaged over the southern hemisphere on 21–22 October 2000. Arrows show direction of N — S transport (C).

angular speed of W — E transport (curve A), total ozone (curve B), and the speed of N — S transport (curve C) averaged over the southern hemisphere. The W — E transport slows down (angular speed decreases, curve A) in middle latitudes and highly accelerates near the pole. Note the formation of a "wave" in curve C (speed of N — S transport). Air, together with ozone, moves towards the hole between 40° and 58° S and back off the hole between 70° and 80°.

A hypothesis of the ozone hole formation in terms of air-flow dynamics

The traced dynamics of stratospheric air has implications for the formation of ozone holes, especially the hole around the south pole.

According to the observed motion of the ozone cloud, stratospheric air moves along meridians and along latitudes. In its motion from the equator to the pole (south pole in our case), it rolls down accelerating poleward because the height of the stratosphere descends away from the equator making a sort of a gradually steepening hill. On the other hand, along-latitude (W — E) motion of air is driven by the Coriolis force. The resulting air motion looks like spindle-like winding at a small step onto the southern and northern hemispheres as far as the poles. The ozone density in the middle latitudes is produced by quite abundant ozone supply from the equator and *in situ* photochemical ozone formation. Ozone near the pole is mostly due to supply from the equator and from the middle latitudes. Its density near the pole is low (Figs. 2 and 5, curve B) as little ozone forms photochemically under the low-angle sunlight and much of the supplied ozone breaks down during the travel which is longer than ninety days, its half-life. The winding of stratospheric air occurs all year round but is especially intense every late winter and early spring both in the northern and southern hemispheres. The reason is that the height of the stratosphere above the Earth's surface is almost invariable at the equator and variable near the pole, being the highest in summer and the lowest in winter when the polar regions are the coldest.

This ozone density distribution and air flow pattern is typical in the absence of ozone holes on both Earth's poles.

Winters on the south pole are always extremely cold, but can be sometimes more or sometimes less cold. In the coldest winters, when the stratosphere is especially low above the surface, the downhill stratospheric air flow speeds up to produce an effect familiar to everyone who ever saw water flowing out of a tub when the plug is out. Once the outflowing water on the surface reaches some speed, it becomes involved in rapid rotation making a "whirlpool" over the hole. The whirlpool is due to the vortex-induced centrifugal force. Something similar occurs with the global motion of stratospheric air flows. As the accelerated stratospheric air rolling downhill gathers enough speed, the centrifugal force pushes it off the pole towards the middle latitudes. The general pattern is that the winding of stratospheric air flows (together with ozone) toward the middle latitudes occurs simultaneously in opposite directions: from the equator and from the pole, the latter motion being driven by the off-pole centrifugal force. This facing migration of air toward the middle latitudes produces a rapidly rotating ring with a very high ozone density.

The stratospheric air connection between the equator and the pole breaks during the life of the ozone hole. Therefore, ozone supply from the equator and the middle latitudes breaks as well and, moreover, the existing near-polar ozone becomes pushed away centrifugally (ozone is heavier than air). As a result, the ozone concentration inside the vortex ring drops dramatically, which is interpreted as an ozone hole. At the same time, the ozone concentration in the ring itself strongly increases.

In summer, when the stratosphere warms up and rises higher above the surface, the downhill rolling of stratospheric air slows down, the whirlpool disappears, the air connection between the middle and high latitudes becomes bridged again, the poleward ozone supply from the middle latitudes resumes, and the hole heals up till the next extremely cold winter on the south pole.

The largest Antarctic ozone hole appears from time to time around the south pole but has no counterpart in the northern hemisphere. Like in the southern hemisphere, ozone in the northern hemisphere moves in spring off the equator to middle and high latitudes. The Coriolis force likewise winds the air flows onto the northern hemisphere from the equator till the pole (of course, winding is in the opposite direction). The ozone content in the middle latitudes is likewise higher than at the equator due to ozone supply and photochemical formation. At the north pole, as well as near the south pole, very little ozone forms photochemically but stratospheric flows bring ozone from the equator and from the middle latitudes. Yet, ozone holes never arise around the north pole, where winters are warmer than at the south pole and the stratosphere is never as low as to provide the critical speed of the downrolling stratospheric air required for a whirlpool.

There is another essential point of difference between stratospheric flows in the southern and northern hemispheres. The W — E air flows in the northern hemisphere are about twice slower than in the southern one (their monthly mean angular speed values are compared in Fig. 6). The reason is obvious. Antarctica is surrounded by oceans with a circumpolar oceanic current. In fact, rotation involves an enormous mass of water and air together. The northern hemisphere, on the contrary, is mostly continental with large mountain provinces at middle latitudes. Friction of air against the rugged land surface prevents the W — E winds at middle latitudes from gathering enough speed.

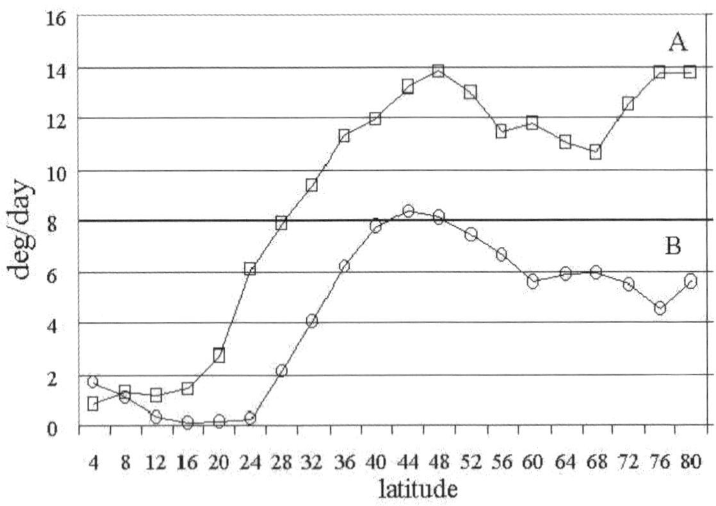

Fig. 6. Monthly mean angular speed of circumpolar vortices in the southern hemisphere between 1 and 30 September 2000 (A) and in the northern hemisphere between 1 and 29 March (B).

The topographic ruggedness in the mountainous continental northern hemisphere produces an air flow pattern at the middle latitudes similar to that of water in a rapid shallow rill with a stony bottom where vortex movements form numerous small bosses and whirlpools on the water surface. The bosses and the whirlpools appear and disappear on the water surface over the respective bosses and pits on the bottom. The water whirlpools in a rill rotate either clockwise or counterclockwise depending on the specific features of bottom topography. For the mid-latitude stratospheric air flows in the northern hemisphere, they are temperature contrasts at the land/sea and plainland/highland interfaces that play the part of topography elements. The quite numerous places of temperature contrasts produce vertical air flows which makes the W — E stratospheric flows of the northern hemisphere twirl in vortices from time to time (of a much smaller scale than over the southern hemisphere). The same air-flow mechanisms maintain lower-ozone areas inside the vortices and higher-ozone rings around them.

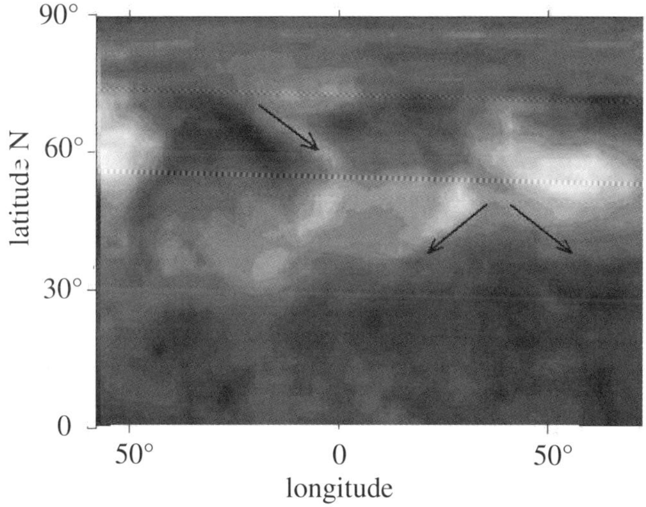

Fig. 7. Satellite image of the ozone layer in the northern hemisphere on 3 September 2001. See three rotating vortices (direction shown by arrows) and a part of another vortex on the left. The vortex on the right is high-ozone (up to 510 DU); on the left there are two low-ozone regions (280–300 DU); the fourth vortex is high-ozone. Zero longitude corresponds to the Greenwich meridian. Light is high total ozone, dark is low total ozone.

The stratospheric air in the polar whirlpool always rotates in one direction determined by Earth's rotation and the mid-latitude flows may rotate either in one or in the opposite direction controlled by the casual pattern of vertical air flows. Vortices rotating in different directions were discovered at first search (Figs. 7, 8).

The dynamics of stratospheric air traced using the new method against the motion of ozone clouds provides a plausible explanation for the formation of the Antarctic ozone hole.

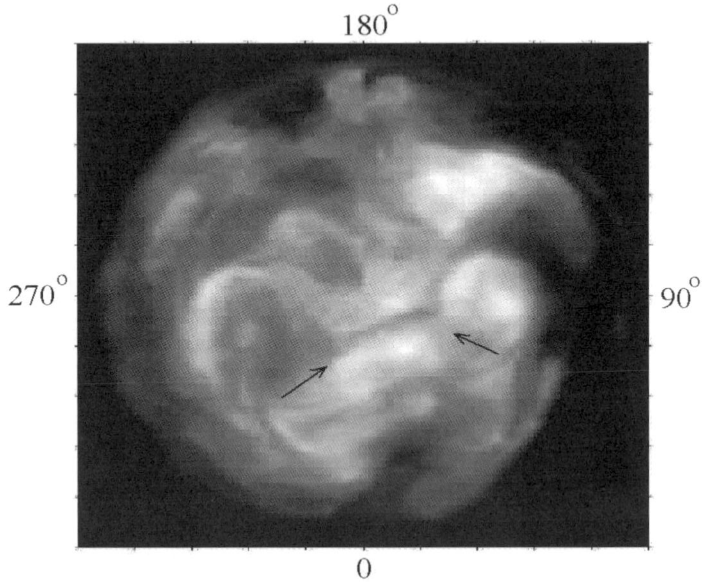

Fig. 8. Satellite image of the ozone layer in the northern hemisphere on 30 April 2003. See a pair of vortices rotating in opposite directions (shown by arrows). The eastern hemisphere vortex corresponds to a high-ozone region and that of the western hemisphere to a low-ozone region (ozone hole). Zero longitude corresponds to the Greenwich meridian. Light is high total ozone (up to 550 DU), dark is low total ozone (260 DU the lowest near the equator).

Its formation is associated with the Antarctic circumpolar vortex, a steady cyclone in the lower stratosphere, from late August to late November. Ozone remains within the natural concentration in summer, autumn, and early winter as it migrates from the equator to the middle latitudes and on to the pole. The Coriolis force causes spindle-like winding of stratospheric air flows (with ozone) onto the two hemispheres as far as the poles. In extremely cold southern hemisphere late winter-early spring seasons, when the stratosphere is very low above the Earth's surface, the ozone supply from the equator breaks at the middle latitudes because a rapidly rotating whirlpool forms over the south pole as the downrolling stratospheric air gathers enough speed. Ozone from inside the whirlpool becomes pushed off centrifugally off the south pole and air flows wind off the pole toward the middle latitudes. As a result, the ozone content near the pole decreases dramatically (an ozone hole appears) and a rapidly rotating high-ozone ring appears in the middle latitudes.

Large ozone holes never arise at the north pole where winters are not cold enough. Small ozone holes may appear in the middle northern latitudes but these are of different origin. Rapid stratospheric W — E air flows respond to temperature contrasts on the earth's surface which cause vertical flows in the troposphere. As a result, stratospheric winds over the vertical flows generate clockwise and counterclockwise vortices. Low-ozone regions (ozone holes) arise inside the vortices, with their characteristic size much smaller than the south pole ozone hole.

The suggested mechanism contradicts neither the dynamic nor the chemical models of the ozone hole formation. There is no reason to believe that ozone depletion would not occur and technological effects would be inoffensive. Yet, the ozone holes are, in our view, primarily produced by natural processes in the terrestrial atmosphere.

We are far from claiming that CFCs and other industrially manufactured gases cause no damage to the ozone layer. However, it remains unclear which exactly are the shares of natural air-flow and man-induced mechanisms. This question inevitably arises with complex natural phenomena, and hasty conclusions are unacceptable.

REFERENCES

Dibble R.R., 1989. In: *Volcanic Hazards. Assessment and Monitoring*, ed. Dibble R.R., Springer Verlag, Berlin-Heidelberg-New York, 536–553.

Farman J., Gardiner B., and Shanklin Farman J. (J), 1985. Large losses of total ozone of Antarctica reveal seasonal ClO_x/NO_x interaction, *Nature*, 315, May 16, 207–210.

Kashkin V.B., Khlebopros R.G., and Kolyada M.N., 2002. Satellite total ozone data as an indicator of stratospheric dynamics: a new interpretation of ozone holes, Institut des Hautes Etudes Scientifiques, Paris, Preprint IHES/M/02/02, 14 pp.

Maduro R. and Schauerhammer R., 1992. *The holes in the ozone scare: The scientific evidence that the sky isn't falling*, 21st Century Science Associates, New York, 356 pp.

Molina M. and Rowland F.S., 1974. Stratospheric sink for chlorofluoromethans: chlorine atomic-analysed destruction of ozone, *Nature*, 249, June 28, 810–812.

Roan S. L., 1989. *Ozone Crisis: The 15-year evolution of a sudden global emergency*, John Wiley & Sons, Inc., New York.

Chapter 6

Closed Ecological Systems and Earth's Biosphere

The salutary adaptive capacity of nature often claimed can no longer keep up the inevitable further technological growth. A solution may come from high-technology biological systems of life support designed on the basis of a solid theoretical background. These systems are tested using manmade closed ecological systems that imitate the Earth's biosphere [*Gitelson et al.*, 2003][a]. A life system is referred to as closed if it recycles all its biological and technological wastes. A fully closed system is tightly isolated from outside but remains energetically open. The closed systems are sustained by the activity of living organisms, mainly plants.

Space vehicles offer an example of workable closed systems, though not yet fully closed. The problems related to people survival, vehicle design, and servicing for space missions miniature the problems of sustainable development in the Earth's biosphere. On the other hand, Earth itself is like a very big space vehicle, and helping its sustainable development, in a sense, belongs to the task of creating a life system for long extraterrestrial missions. Of course, the operation of neither the Earth itself nor a manned space flight can be all put into formulas and figures, as any complex system. The point is which parameters to monitor and which processes to close and to predict.

Conservation of resources for the coming generations is just a political catchword but reducing consumable resources is a vital

[a] All data below concerning manmade closed ecosystems are given according to this book and references therein.

objective of astronauts' survival as it saves the costly, difficult, and unsafe supply from Earth. The relatively small spacecraft can change its inner air composition in a few days while the same change in the Earth's atmosphere would take ages. Therefore, the life systems for spacecraft and the manmade biospheres, their terrestrial test samples, are a sort of an "ecological time machine" allowing prediction for the planet future. Below we consider energy, heat, air, and food cycles in manned space vehicles as a model of the corresponding processes in the biosphere. Another reason to invoke space-related issues in the context of high-technology manmade closed ecosystems is that the technologies for space applications are commonly recognized as the most advanced ones.

Energy in space vehicles

Energy use is a key point in the global environmental crisis as the type of fuels essentially controls the structure of economy and thus its impact on environment.

Space vehicles use chemical, solar, and nuclear power, each applied where it is indispensable or more gaining. The choice of power supply depends on whether the flight covers the low Earth orbit, interplanetary (from Mercury to the asteroid belt) or intergalactic space, outside the asteroids. The space between the Sun and Mercury remains so far beyond the coverage because of the high solar gravitation and radiation.

The use of chemical energy associated with oxidation (firing) is restricted to taking off and landing which requires high peak power[b]. So intergalactic space vehicles have started from and landed onto the Earth, while other celestial bodies have been explored by small stations, likewise moving by chemical fuel. Thus chemical energy has been mostly used within the low Earth orbit space and is also necessary for landing on the Moon and other planets. In manned spacecraft, harmful combustion materials are normally removed outside the life systems, though can penetrate inside in emergency. American astronauts

[b] Peak power is a great amount of energy released in short time.

happened to get poisoned because of seal failure, fortunately nobody was killed. That accident prompted designers to take special measures for isolating the living module from the fuel-energy systems.

Chemical (fuel) energy cannot compete with solar energy in space, especially in the zone of high sun radiation in interplanetary space. Indeed, a solar array about 1 m^2 in surface area and about 10 kg in weight can run for tens of years producing an energy of 100 W. Firing 4 kg kerosene with 6 kg oxygen, a total of 10 kg chemicals (on Earth oxygen freely comes from air but has to be brought in from Earth to the orbit), yields about 8000 kcal of heat equivalent to at most 12000 kJ of electric energy and to the energy produced by a 100 W solar battery in eight days. The same ten kilo (including oxygen) of the most efficient hydrogen fuel would yield this energy in fifteen days. Therefore, fuel energy obviously loses the competition in the conditions of space, and (except for electric batteries) is never supposed to be used in open space where no peak power is needed. The obvious advantage of solar energy sets a good example for terrestrial applications.

Solar energy can maintain life systems and, moreover, was projected to drive ion engines in extra-planetary vehicles. At present ion engines use electric energy to accelerate heavy particles (say, ions of cesium or mercury) ejected back from rockets, like a gas jet in a chemical jet engine. The energy growth in a rocket is proportional to square velocity of the ion jet[c]. The limited amount of ion fuel commands careful choice of the optimum velocity of ions. Going to Mars which takes about six months one way requires that ions move at 40 km/s. The needed amount of energy can be provided by a large sail-like solar array. The sails open up in space to make an ion-engine spacecraft surprisingly similar to a sail boat, but its sails meet no air resistance. The today's technology would allow an about 120 kg weight per 1 kW of solar array equipped with all necessary facilities; future technologies may save the weight considerably.

Solar arrays have been so far applied to maintain life and mechanical systems of spacecraft. Note that the array efficiency varies with distance from the Sun (solar radiation is inversely proportional to square distance

[c] Kinetic energy is given by $E = mv^2/2$.

from the Sun). A solar array is thus twice stronger on Venus's orbit and almost twice weaker on Mars's orbit than on the Earth; as far as at the Jupiter's orbit, the efficiency is twenty times as low. Therefore, the use of solar energy is restricted to interplanetary space.

Nuclear energy yields high specific power per engine unit weight but requires a sophisticated cooling system. It is profitable in far space in the deficit of solar energy. *Voyager* used a nuclear energy source for distant travels. There were attempts to use nuclear energy for Earth-orbiting satellites but this was found risky. Indeed, a nuclear engine is inoffensive in space unless a satellite left on the orbit eventually descends back on Earth and collapses spreading nuclear fuel through the atmosphere, even if soft landing is tried. Fortunately, the reported accidents with American nuclear sources (e.g., *Apollo 13*) escaped polluting the environment. In the emergency of *Kosmos 1402*, the Soviet satellite, fuel dissipated over the land when the satellite entered the atmosphere. Yet, a few kilos of radioactive matter spread evenly over the global surface will cause no dramatic effect on the rather high natural background.

The problem of energy supply in spacecraft is related to the problem of air heating inside. It is undesirable to supply more energy than needed because all input energy eventually converts to heat to be taken out of the system. Radiation into space (cooling) is the only way to remove excess heat from any spacecraft, big or small (and, hence, also from the planet Earth, an enormous space vehicle). According to the Stefan–Boltzmann law, radiation of any ideal black body[d] is proportional to the fourth power of its Kelvin temperature (counted from –273°C). The normal air temperature we live in, about 300K, is maintained inside spacecraft, and a 1% of excess heat energy causes a temperature increase of 0.75°.

To be radiated to outer space, heat is conveyed to spacecraft walls by special cooling systems which constitute a considerable portion of all systems and weigh 100 – 200 kg per 1 kW of removed heat, heavier than in solar array. In principle, there is a possibility to condense excess heat (which requires additional energy) and take it to special heated radiators

[d] Ideal black body absorbs all received radiation, which is approximately the case of all space objects of our interest.

where cooling is much more efficient the radiation being higher at higher temperature. This is unrealistic on Earth but has been succeeded lately for spacecraft. Japanese reported they reduced the weight of cooling systems to 2 kg per 1 kW of rejected heat.

A cooling system is not always able to maintain the optimum air temperature in spacecraft as temperature depends on energy input and other factors. For example, inner air temperature fell to +3°C because of emergency depressurization during the *Apollo 13* mission, and people had to use the warmer but still cold (+11°C) lunar module for life support. Another example is the sunshade emergency on *Skylab* that caused temperature rise to as high as +60°C.

Life support systems

The human lungs are perfectly adapted to the natural gas mixture they draw oxygen from (78 vol. % nitrogen, 21% oxygen, 1% argon, 0.036% carbon dioxide (dry air), and minor contents of other gases). Man's survival depends on free oxygen supplied by plants. The atmosphere of other planets, in the absence of plants, contains no free oxygen as it quickly binds with other chemicals. People can live with at least 10% atmospheric oxygen at normal pressure; people accustomed to oxygen deficit, e.g., mountain dwellers, feel well with slightly lower percentages than all others, and can go with much less oxygen for a short time; untrained people become sluggish already at a moderate oxygen decrease. Pure oxygen is suitable for breathing as well but for a short time; its healthiest percentage is about 21%.

Carbon dioxide is needed in a small amount to regulate respiration. People never care about its least concentration as they exhale enough carbon dioxide but the upper limit is essential. Already 5% CO_2 is fatal, 0.3% causes notable change to the human organism and can be assumed as the maximum allowable limit, though people had lived safely with 0.8 – 1% CO_2 for six months in the closed system of *BIOS 3* (see below).

Plants, on the contrary, consume carbon dioxide and can live between 0.01% and 5% (Fig. 1).

The qualitative image of Fig. 1 shows that plants can exist with much broader percent ranges of atmospheric gases but people have to care about the air composition good for both humans and vegetation which gives them oxygen and food.

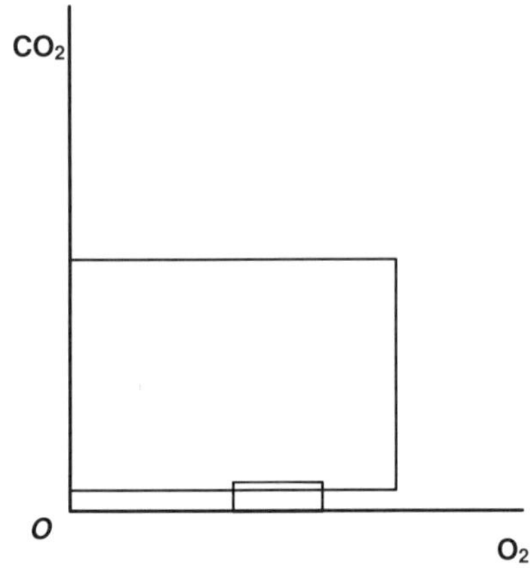

Fig. 1. Vital percentages of gases for humans (small box) and plants (large box). Overlap corresponds to the natural living conditions for people.

The Earth's biosphere is sustained by two key processes: photosynthesis in which plants synthesize nutrients (carbohydrates, lipids, and proteins) using sunlight and the contrary process of decay provided by animals, fungi, and bacteria.

For carbohydrates the two processes are:

$$CO_2 + H_2O <=> \text{(carbohydrates)} + O_2.$$

Fructose, the primary product of photosynthesis, forms by

$$6CO_2 + 12H_2O <=> C_6H_{12}O_6 + 6O_2 + 6H_2O^{e}.$$

---

[e] The water molecules in the equation were not cancelled to highlight the fact that oxygen forms from water rather than from carbon dioxide.

Carbohydrates include cellulose (largely synthesized in nature but consumed almost uniquely by fungi), amylum, glycogen, honey composed of a carbohydrate mixture, sugar, etc.

Metabolic processes also involve a lesser component of lipid cycles:

$$CO_2 + H_2O <=> (\text{lipid}) + O_2.$$

Lipids have a diverse chemical structure. Many known lipids are based on glycerin and fat acids but living organisms rather need the lipids that enter cell membranes.

With $C_{16}H_{32}O_2$ taken as the standard lipid compound, the chemical balance is given by

$$16CO_2 + 16H_2O <=> C_{16}H_{32}O_2 + 23O_2.$$

A protein cycle requires additional nitrogen-bearing compounds:

$$(NH_4)_2CO + CO_2 + H_2O <=> (\text{protein}) + O_2 .$$

The greatest amount of proteins is found in rapidly growing organisms (up to 70% in some bacteria). Proteins catalyze chemical reactions in organisms and are thus crucial for life. Proteins define the specificity of organisms and the constituent individual cells.

With C_4H_5ON taken as the standard protein and $C_2H_6O_2N_2$ as a nitrogen-bearing compound, the respective balance is given by

$$C_2H_6N_2 + 6CO_2 + 2H_2O <=> 2C_4H_5ON + 7O_2.$$

Of course, metabolic cycles involve other compounds besides carbohydrates, lipids, and proteins and almost all chemical elements besides carbon, oxygen, hydrogen, and nitrogen, but their percentages are vanishing and can be neglected in the approximate models we discuss.

According to the above equation for carbohydrates, plants produce six oxygen molecules per six carbon dioxide molecules they consume.

The same uptake-release balance is found in all processes of carbohydrate production by plants. In synthesis of lipids the number of produced oxygens is about 1.5 times that of consumed CO_2 molecules: 16 CO_2 per 23 O_2 (see the above equation). The proportion is roughly the same in other reactions of lipid building. Thus, synthesis of lipids releases 1.5 times more oxygens than synthesis of carbohydrates, given the equal number of CO_2 molecules. All plants produce both carbohydrates and lipids but in different proportions, which is essential for closed life systems (see below).

Photosynthetic reactions are biochemically similar in all plants, from the lower microscopic algae to the higher cultivated plants, but decay provided by the activity of different organisms occurs in different ways. Kangaroo rat that lives in deserts and eats dry grain never drinks but uses metabolic water released by food decay in its organism. Cellulose-rich tree biomass cannot be assimilated by animals and is decayed by fungi.

The proportion of carbohydrates, proteins and lipids differs in diets of different peoples worldwide. The diet of Oceanians is poor in animal and even vegetable proteins but is dominated by carbohydrates, while the diet in the Arctic regions, where people lived on hunting and fishing until recently, includes mostly lipids and proteins. Astronauts use the European diet containing 60–65% carbohydrates, 20–25% proteins, and 10–20% lipids.

In the biosphere, everything an organism produces, and eventually the organism itself, become food for other organisms, and solar energy at the input maintains the cycles of all matter. It is this metabolic vortex that biochemistry refers to as life. The same atoms are used and reused in metabolic cycles and enter diverse compounds in various living organisms. If this natural chemical cycle were broken, life would end in a historically short time from 40–60 to at most 1500–2000 years, depending on the break point. Thus life on our home planet exists due to the globally closed chemical cycle[f], which is currently being disturbed by the technological activity of man (see Chapter 4).

---

[f] Yet, the closure is not perfectly full. There are organic wastes dropped out of the biochemical cycles which left geologically significant signature for the time since the onset of biota.

The activity of living organisms contributes to contamination and cleaning of water and air. Cleaning obviously goes by energy supplied to organisms from outside. Man takes about 30 kg air with 6 kg oxygen per day through his lungs to draw the necessary 0.6 kg oxygen; the exhaled air is unfit for breathing without cleaning. Moreover, man, a consumer of water, though he is a water producer biochemically, takes 4–5 kg pure water through the organism; the water cycled in a human organism likewise needs cleaning before reusing. Plants, on the contrary, are biochemical water consumers but evaporate a great amount of water in their life cycles and use most of the received sun energy for water cleaning.

Mass exchange in the biosphere occurs by chemical cycles with participation of living organisms and by physical cycles of which the main are evaporation from the oceans and precipitation onto the land i.e., cleaning and transport of water.

The chemical and physical biospheric cycles have to be simulated in the systems for life support outside the Earth. The conditions of the Earth's biosphere can be maintained using plants, animals, and microbial organisms, and, additionally, various technological systems. Some models of closed life systems have been already tested on the Earth.

The available life support systems are yet not fully closed yet though are getting ever more autonomous. The first space vehicles supplied water and oxygen, used them separately and never recycled. The today's practice on the *MIR* station is to recycle water and air; shuttles deliver drinking water while oxygen is extracted therefrom by electrolysis. The liquid wastes are collected and the exhaled air is condensed in the cooling systems. The amount of recycled water even exceeds the original amount because more water is released from food, even if it is dry. Carbohydrates form of carbon dioxide and water — which is evident from their name — and decompose back in the human organism to release carbon dioxide and the metabolic water. Therefore, advanced recycling technology can cancel water delivery to the orbit.

Thus each oxygen atom from the supplied water is used and reused. An astronaut can get one with water to inhale the same atom in a while with air; recycled in the organism, it turns back to water. People on the

MIR station spend oxygen more sparingly than the kangaroo rat: they produce oxygen themselves while the animal takes it freely from air. This is quite natural as the space conditions are harsher than in any desert. A minor percentage of O atoms are evacuated from the station with exhaled CO_2 that remained beyond recycling.

The quality of air inside a spacecraft is a subject of special care. The quality of inner air depends on the contents of volatiles released by people and utensils they use, from the navigation instruments and other things delivered to the orbit. The best way to maintain high air quality is to use plants which surpass any manmade air purifier. However, the higher the quality demand the bigger and heavier biological and mechanical recycling systems are required.

The optimum configuration of life support systems and the strategy of a space mission as a whole depend on its duration. A light life system is more beneficial for a short stay when more supplies are acceptable as they won't be too heavy anyway. Long missions require saving life systems which allow a quite big starting mass of the supplies. Reducing the mass of systems and supplies is important also for the reasons of safety as excess load on the engine enhances the emergency risk. Highly closed systems are the best mass saving systems for long space missions. If life systems kept their configuration invariable, they would grow heavier with increasing supplies. In practice their configuration changes repeatedly during a mission until it reaches the maximum mass closure.

Fully closed life systems are applicable in short flights as well but their lower mass decreases their reliability. Gagarin's spacecraft did not need a closed system because it was designed for a few days stay in orbit. A Mars mission would go better with a more closed life system (see Fig. 2); the one like in Gagarin's spacecraft would require great supplies and would break in a short while because of the lack of spares. In Fig. 2 the starting mass M (recycling and maintenance systems plus supplies) is plotted against the mission duration t for three versions of life support systems.

The mass M is the least in a but the system a is open and commands greater supplies than b. The system b has a greater starting mass but allows more closure thus saving the food supplies. Finally, c is a fully closed system which is heavier but needs no supplies, as food is recycled and the mechanical systems are supposed to be repaired instead of using spares. The starting mass is thus not proportional to the mission duration. The case of Fig. 2 is that the open system *a* is better for short flights until the curves *a* and *b* cross at point 1; *b* is better between points 1 and 2 (semiclosed system), and, finally, the fully closed system *c* after point 2 is preferable for long missions.

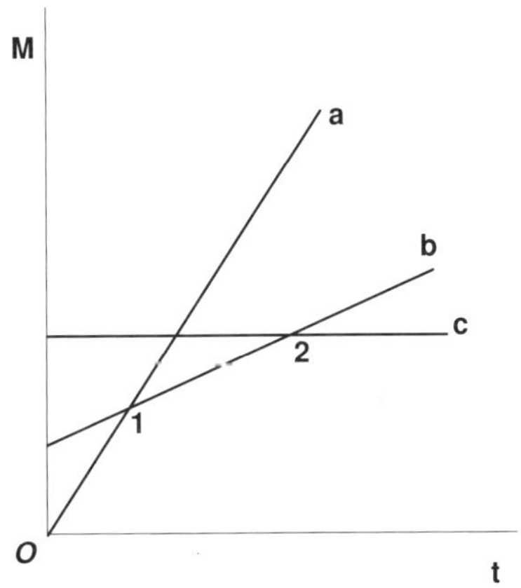

Fig. 2. Starting mass (*M*) of spacecraft (recycling and maintenance systems plus supplies) as a function of mission duration (*t*) for three versions of life support systems.

The discussion of closed ecological systems has implications for theoretical ecology as well as for practical issues of life support in space missions and in extreme conditions of terrestrial environments.

Designing space life systems should proceed from the mission type and duration. The weight of the system is of lesser importance on a long-run Moon station where supplies are brought in only once and Moon ground is used for various needs. A station maintained yearly by shuttles requires a lower mass of supplies.

Another important problem is to mitigate the life risk for people. The main jeopardy comes from emergency in taking-off and landing: the available statistics predicts about 2% probability for an emergency at each taking-off and the ensuing landing. In this respect long missions of about three years are less risky. However, a too long stay is dangerous because of tiredness which increases the risk of mistakes and health problems. Space stations should offer highly comfortable conditions to prevent people from getting tired and allow a longer stay. The optimum strategy will be to differentiate between the safety of a mission as a whole and the safety of each individual astronaut. When on the station, an astronaut gets tired mentally and physically and takes more risk of getting sick on return; therefore, he has his own reasons to come back sooner. On the other hand, the long stay in space and the less frequent crew change reduce the total risk for the mission as a whole (as in the case of the *MIR* station).

Manmade closed biospheres

Life support systems used so far in space vehicles are not fully closed. The first closed life systems, the manmade biospheres, were tested in land experiments. The simplest non-manned biospheres were just ecosystems sealed in vessels containing algae and bacteria or algae and fungi. Life died in some vessels but some ecosystems turned viable and available for studies. Nobody could predict whether the community survives or not and which form it will take.

First manned biosphere rest module experiments were run in the mid 1970s at the Institute of Biophysics in Krasnoyarsk as part of the Soviet space programs. The best known closed ecosystem suitable for human life is *BIOS 3*, designed in the late 1960s-early 1970s, where three men lived safely for six months. By its configuration, *BIOS 3* is a prototype of a life system for a moon station. *BIOS 3* imitated the Earth's biosphere in that oxygen, water, and food were produced by plants. The system was airtight (though there was minor leakage) and energetically open (electrical power was supplied from outside); cooling was provided by running water and communication was available by telephone. People spent about two hours a day for life support activities and had enough time for research.

The ecosystem included plants which gave oxygen and vegetal food and bacteria that belong to the common microflora in humans and plants.

Plants maintained the oxygen cycle, and the air quality remained good all the experiment long.

An ideal chemically closed system would imply, for example, a single human-complementary recycler plant to secure necessary food and oxygen and recycle all wastes (Fig. 3).

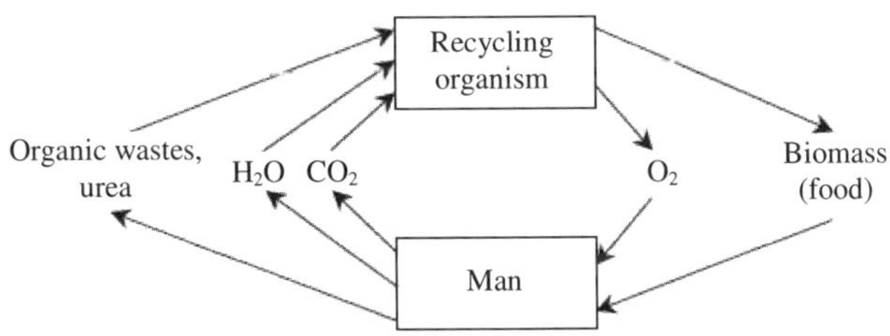

Fig. 3. An ideal chemically closed life support system with a single recycling organism.

For the lack of such a plant in nature, several plants are used to approach fully closed systems. The systems that have been tested so far in theoretical calculations and laboratory experiments included a few recycler organisms (microscopic algae, hydrogen-producing bacteria, and higher plants). A fitting closed life system may look as in Fig. 4.

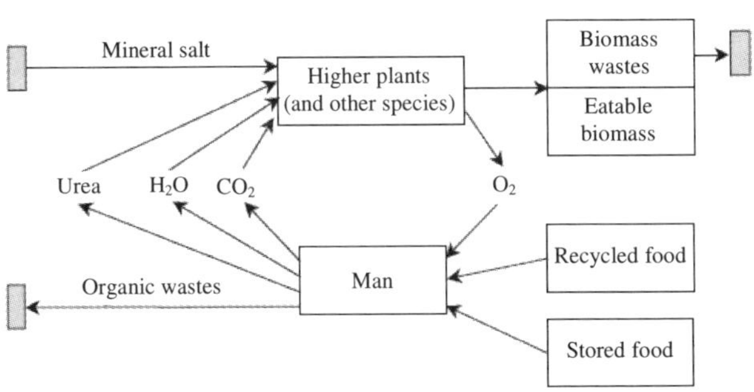

Fig. 4. A closed life system with several recycling agents.

Recycling in *BIOS 3* was sustained by higher plants. In principle, biological life support systems can include domestic animals. There was an attempt to adapt a she-goat to a closed system but it did not feel well, possibly, it suffered from the lack of motion. Other experiments tested the adaptation of molluscs and fishes. Keeping animals requires ten times the amount of energy needed for growing plants, and obtaining food through these mediators is costly. It goes well in nature by the free energy of sunlight whereas energy supply in space is limited by the available instrumental facilities.

Food demand for *BIOS 3* was estimated as a balance of four basic elements: carbon (C), oxygen (O), hydrogen (H), and nitrogen (N) which together make up 98% elements involved in the biochemical cycle; O, CO_2 and H_2O make up 75%. Food production by plants was calculated so that it satisfy the demand with the selected normal diet. People recycled almost all compounds to turn them back to plants. The inedible remains of plants (straw) caused a problem. The crew of *BIOS 3* dried straw and put it aside. Some experiments tested burning straw to get carbon hydroxide for feeding plants but no success was achieved. Another solution is to oxidize straw under high pressure to obtain compounds usable by plants. Although in general the plants received from people roughly the same quantities of the same compounds they gave, the *BIOS 3* experiment did not achieve full recycling of plant wastes[g].

Plants cannot be the only components of a human diet for a long time because their balance of aminoacids is different from that needed for people. The diets of all peoples, even if mostly vegetarian, include some meat or milk food. Therefore, following medical requirements for the nutritional balance, the crew of *BIOS 3* used supplies of vacuum-dried meat and other animal food (their diet remained virtually the same as in the everyday life), which obviously made the closure incomplete. Otherwise, the deficit of aminoacids in a vegetarian diet can be paid off by 30-50 g aminoacid extracts per day, which is quite suitable for the conditions of short space missions.

[g] Later tests offered better ways of recycling for feeding plants.

Thus the food cycle in *BIOS 3* was not fully closed as it did not recycle some vegetal biomass and some human wastes, and supplied about 50% of the food. Given the air closure (see below), a totally closed system implies, according to the conservation law, that the food supplies and the non-recycled wastes were balanced in the amount of chemicals: the added mass of elements has to equal that evacuated from the system.

Air closure is another challenge. The air balance of *BIOS 3* was calculated for carbon dioxide (CO_2) and oxygen (O_2). When exchanging chemicals, people and plants give and receive the same quantity of atoms but not necessarily the same quantity of molecules the atoms build. The ratio of the number of consumed CO_2 molecules to the number of produced O_2 molecules in plants is called *assimilation quotient*. In humans this quotient ranges from 0.83 to 0.86, depending on the diet. Different plant species have different assimilation quotients. Wheat has 0.92–0.94 and thus cannot sustain air balance with man. Oil plants have an assimilation quotient lower than wheat, as synthesis of lipids releases 1.5 times more O_2 molecules per one CO_2 molecule than synthesis of carbohydrates. Some oil plants have their assimilation quotient lower than that of humans. Therefore, to maintain the air balance, one has to approach the human assimilation quotient by fitting the proportion of wheat, vegetables, and oil plants. This discovery essential for air closure between humans and plants was put into practice in Krasnoyarsk (but nowhere else, as far as we know). The crew of *BIOS 3* used chufa-nut (*Cyperus esculentus*), an oil plant from Central Asia, that maintained the air balance and besides gave indispensable vegetal fat.

The gas balance modeling for *BIOS 3* faced a problem of microbial organisms, especially the poorly known soil bacteria. Soil bacteria, which in nature recycle organic wastes, did not fit any calculation. Therefore, soil was withdrawn from the system and plants were supposed to grow in water as hydroponic culture[h]. Moreover, special measures were taken to combat the pathogenic flora, and the system included only the bacteria that are the common symbiotic companions of man. Their presence outside the human organism is inoffensive and could be

---

[h] Besides the air closure, soilless agriculture saves the mass which is essential for long space missions.

neglected in the models. The bacteria of plants were likewise left beyond the predicted air budget.

Thus, the air budget included a few well studied species of plants specially selected for space missions, namely wheat, chufa-nut, and vegetables. They fed hydroponically from a nutrient solution and grew without a "night rest". Wheat lighted with high-power lamps for twenty four hours a day cropped as early as in two months. The plants occupied a surface area of fourteen square meters per person (that is to see again how the advanced technology can reduce agriculture areas).

One project of a moon station with its life system similar to that in *BIOS 3* implied a crew of twenty persons and a densely packed plant unit made as a cylinder 2 m high and 8 m in diameter to ensure the necessary amount of oxygen, vegetal food, and clear water.

Due to the air closure, the composition of inner air in *BIOS 3* needed no detailed monitoring. The food cycle would go by itself if people had a vegetarian diet corrected with minor amounts of animal aminoacids. This configuration would require twenty five square meters of plants per person; the non-recycled plant wastes would be burnt or otherwise processed.

Being closed in air and semiclosed in food, *BIOS 3* was energetically open. Energy was supplied from outside to maintain illumination twenty four hours a day at the level necessary for photosynthesis, about 3 kW for full life support of one person. In nature this amount of light can be obtained from a surface of about forty square meters in the interplanetary space and from two hundred square meters on Earth, where illumination depends on latitude. Excess heat was removed from the system using the common river water to cool the modules from outside, whereby water from the air inside condensed on the inner walls. Of course, this way of cooling is inapplicable in space missions.

Of many events during the *BIOS* experiments, it is pertinent to mention a miniature "environmental disaster". Today's environment scientists are very much concerned about ozone holes. Ozone holes are breaks in the thin ozone shield in the upper atmosphere that protects the Earth from hard UV radiation (for more detail of ozone holes see Chapter 5). The lamps used in *BIOS 3* emitted UV rays besides the visible light, like the Sun, and were coated by glass to screen the UV radiation. Once the glass coating of a lamp cracked. The crack, an equivalent of an ozone hole, remained open for a few hours and the released ultraviolet killed a part of plants, which disturbed the CO_2 balance. The experimenters replaced the coating (repaired the "ozone hole"), planted new plants and waited what would happen: whether the plants reach the stage of active photosynthesis or the concentration of carbon dioxide surpasses the acceptable level. Eventually, things turned good: the plants saved the situation and CO_2 started to fall back in ten days.

The *BIOS* system was different from other similar modules tested later in Moscow and in the US. Unlike *BIOS*, the Moscow systems failed to achieve full air closure. In the American system, the assimilation quotient of plants exceeded 0.9, and the excess CO_2 was removed by chemical absorption and the absorber was rejected.

Before building the Krasnoyarsk biological life support systems, their operation was scrutinized in theoretical modeling, though it was not obvious *a priori* whether everything can be predicted in detail. The tests of the predictive models during the *BIOS 3* and other similar experiments allowed a better understanding the Earth's global ecology. Thus, the *BIOS* program, although originally designed for the needs of space exploration, had important biological implications beyond the imposed limits.

An approach different from the scientific and engineering background of the Krasnoyarsk biospheric experiments was proposed and tested in the *Biosphere 2* project founded by John Allen and his team. *Biosphere 2* was not supposed to simulate space life systems and ran beyond space programs, though its value as a prototype for permanent life-habitats on suitable locations in space was commonly appreciated.

Biosphere 2 was designed, built and operated by Space Biospheres Ventures, the project was organized as a private venture capital endeavor and cost $162 million. The designers proceeded from a philosophical idea that a manmade biosphere compositionally similar to Biosphere 1, that of the Earth, should be a self-sustaining self-organizing system suitable as a habitat for man.

The world largest closed ecological system of *Biosphere 2* covers 3.15 acres near Oracle, Arizona. The glass and spaceframe structure measures seven million cubic feet in volume (three thousand times that of *BIOS 3*) and stretches up to 85 feet at its highest point. Its steel struts are covered with a finish that insures against corrosion from inside or outside the biosphere. All air, water and nutrient cycles were completely closed and recycled within the system. Though highly airtight, the system was not fully isolated: The exchange of *Biosphere 2*'s air with the Earth's air (a leak rate) was under 10 percent a year.

The energy of sunlight was used by plants for photosynthesis and got eventually converted into heat. Mechanical systems assisted the heating, cooling, and air and water circulation. A separate system condensed water from the atmosphere to supply human drinking water and water evaporation from the ocean was condensed for drinking water and for return to the stream and rainforest. In addition, to prevent the system from imploding under enormous pressures as the air inside expands and contracts depending on the outside air temperature, *Biosphere 2* got a pair of "lungs", or variable expansion chambers, connected to the main structure via underground air tunnels. The pressure difference was kept up by a synthetic rubber membrane with a circular metal top moving freely up and down on a cushion of air.

Scientists created seven complete ecosystems or biomes that mirrored those of Earth. The systems included an ocean, a desert, a savannah, a tropical rainforest, a marsh, an area of intensive agriculture and a human habitat. Each different biome was built from scratch — with carefully selected soils, water and plant and animal life, collected from all over the world. *Biosphere 2* sustained high biodiversity with approximately 3,800 living species.

One of the most important accomplishments for *Biosphere 2* was the agriculture which ran very smoothly and yielded high production. The biospherians grew, harvested and processed their food while keeping the soil highly fertile and using non-polluting pest control methods. Though quite small (half-acre of land), the farm contributed to the recycling of material more than the "wild" biomes, such as, say, rainforest. Therefore, the sanitary impact of agriculture on planet's air can compete with that of the pristine forests. However, the latter are irreplaceable anyway as wood cellulose and lignin remain non-decayed for ages which reduces the concentration of atmospheric carbon dioxide.

The first crew of *Biosphere 2* consisted of eight researchers, known as "biospherians," four men and four women, who safely spent 24 months within the glass walls.

The two projects of *BIOS 3* and *Biosphere 2* are fairly different, not only in the scale of the experiment. Prior to the onset of the *Biosphere 2* experiment, the two approaches were discussed at the Second International Workshop on Closed Ecological Systems held at Krasnoyarsk in 1989. The basic difference was that the founders of *BIOS 3* proceeded from theoretical predictions for the performance of all its units, i.e., the system was built as a "machine with biological blocks" with the Man responsible for their operation, whereas the creators of *Biosphere 2* mostly relied upon the self-organization of biological systems within it.

In spite of its many valuable accomplishments, the *Biosphere 2* experiment has been recognized to be generally a failure in the sense that it has never achieved the expected self-running sustainable system to offer a favorable or at least suitable human habitat. This is an argument against the anti-technological attitudes often expressed in environmental discussions. Indeed, high-technology life systems have been successfully operated while the project which anticipated self-organization of natural processes did not work properly. The oxygen drop and carbon dioxide increase reported in the beginning of the *Biosphere 2* experiment may have been caused by unpredictable activity of soil bacteria. Figure 1 showed the subtle balance of oxygen and carbon dioxide allowing coexistence of humans and plants. An ecological system where this balance (fortunately sustained by the Earth's biosphere) would be disturbed, if survives, may become uninhabitable for man. Other causes behind the air disturbance in *Biosphere 2* may be that some species able to control the air composition became extinct; moreover, outbreaks of some pests devastated a part of crops. Although oxygen was twice pumped in from the outside, its low concentration eventually prevented people from further stay inside the system.

Nevertheless, the *Biosphere 2* experiment was especially important for having proved something different from what was originally meant. Namely, it demonstrated all the fault of hoping for self-organization of natural processes and neglecting the intellectual efforts, as is typical of the "Green" thinking.

On the other hand, the *Biosphere 2* system has become an unmatched laboratory for mitigation of environmental disasters. People should learn from liquidation of the consequences of the disaster in *Biosphere 2* how much effort the repairing of the damaged Earth's biosphere would require. Therefore, the failure of the *Biosphere 2* experiment is paradoxically its success.

Planet Earth as spacecraft

Environment is the most vulnerable point in the today's crisis of civilization for the global environmental disaster is a very likely way in which our civilization may collapse. There have appeared various projects of an environmentally safe future which actually fall into three main groups: (i) a stationary civilization, (ii) a closed civilization, and (iii) Earth as spacecraft.

1. *Stationary civilization*

The approach of *stationary civilization* suggests to stop the technological expansion and to leave the civilization where it had arrived by the 1950s.

This project is extremely dangerous in terms of conservation. The question is whether the biosphere in its today's state can sustain equilibrium if the nature use remains at the same or even at some previous level. People have already launched the processes they can no longer control, and nature won't necessarily return to its previous state by itself if left alone. It is unclear whether nature would regain its equilibrium without the help of man if the threshold of self-regulation has been surpassed. The experience of *Biosphere 2* warns of how unwise it is to rely upon mere "intuition" and "common sense". What is really needed is scientifically grounded prediction continuously tested against empirical data.

Stabilization of the world civilization at the highest level it has achieved, commonly attributed to that of the US, will inevitably bring the world to the economic and environmental collapse at the available state of technology.

Moreover, the idea of a stationary civilization encourages an attitude of stagnation.

2. *Closed civilization*

The *civilization of closed cycles* implies almost no contact with nature rather than its exploitation. According to this project, the sphere of civilization — called noosphere not quite following Vernadsky's concept — is strictly separated from the biosphere; biosphere is left untouched being preserved as a park for guided tours and safe investigation; noosphere is divided into closed technological cycles within which all wastes are recycled and used for further production; people are supposed to have a closed metabolism.

The closed cycles obviously need energy from outside, much more than in the normal technological processes, and will be thus energetically open. They may use the environment-friendly and almost free solar energy, which may be quite possible in the future. Unless noosphere is divided into fully closed cycles, the exchange of products between them will break the closure. Actually, noosphere as a whole will become an incredibly sophisticated enterprise. Predictions for its operation will face the difficulties common in the modeling of complex systems in which only few of the functions are liable to formalization. This cancels the advantage of predictability attributed to closed systems. Finally, the idea of a closed civilization likewise implies stagnation, as well as the previous approach. At this point both ideas appear to contradict the nature of man as an active being unable to stop in its evolution.

3. *Planet Earth as a spacecraft*

Vernadsky interpreted Earth itself as a closed system. The operation of its entire system cannot be predicted theoretically but a number of selected parameters can be monitored, especially the balance of mass and energy. In this and the previous projects of a closed civilization and an Earth as a spacecraft it is urgent to prevent Earth from heating and refuse carbon fuel to mitigate the greenhouse effect. The operation of the system can run in accordance with short-term predictions tested against measurements; more parameters can be taken for monitoring if necessary. Shortly speaking, in this project, as well as in the project of a closed civilization, man charges himself with the control over the planet Earth.

Sparing consumption of materials and energy that preserves the biospheric equilibrium may allow colonization of space where there is enough space for unbounded development.

REFERENCES

Gitelson I.I., Lisovsky G.M. and MacElroy R.D., 2003. *Manmade closed ecological systems*, London, Taylor & Francis, 402 pp.

Chapter 7

Environmental Damage

Modern man and environment

The economic activity of man has various environmental consequences, often damaging. Environmental damage can be sometimes hard to estimate, especially when it affects unique natural sites. Aggressive nature use stirs up public protests. People are very much concerned about extinction of rare or relict species — even though these may have no evident economic value — or about degradation of rivers, lakes or seas. Public outrage is in some cases efficient. Strict legal measures set in response to public claims saved Great Lakes in America which were about perishing in the 1960s. Now they are good for recreation, though attempts to recover some valuable fish species have had no success yet.

Environmental risks are especially serious in totalitarian states where destructive nature use is imposed by the authorities and any attempts of public protests are quelled. This was, for example, the story of building a paper mill on the shore of Lake Baikal or the degradation of the Aral Sea in Russia.

Gradual pollution of air, water, and soil is less notable being an everyday experience, and the public response to the danger can come too late. However, combating the everyday pollution is by no means hopeless. For example, air in American cities has been essentially improving due to new car filters, etc.

Living in the market economy (see Chapters 9–14), people are forgetting their ancestors' wisdom of nature care. Prehistoric hunting tribes practiced rigorous restrictions, often seasonal, on killing certain animal species. Those restrictions (taboo) were supported by religious rites and often helped conservation of the tabooed species. Many tribes lived on hunting but never threatened the basic animal populations. The American Indians hunted bisons, but they were Whites who exterminated the million-strong bison herds without any economic need. Later on, in the time of agriculture, people learned to maintain soil fertility for centuries giving back to the ground the substances they withdrew and sparing the related ecosystems. However, the traditional wisdom of peasants gave way to the careless attitudes of the modern industrial society, as was mentioned, for instance, by Konrad Lorenz, the greatest biologist of our time.

The destructive impact of civilization on environment outrages many intelligent people but most people follow the motivation of their immediate interests (see Chapter 14). Even ensuring the survival conditions requires restraining producers who would not stop at throwing wastes over the heads of their neighbors. Environmental restrictions can be reasonable and efficient only if they have solid scientific background, specifically the knowledge of the dynamics of environmental damage.

Dynamics of environmental damage

We model the dynamics of environmental damage applying for simplicity the term "pollution" to its any kind (release of aggressive chemicals, either natural or synthetic, deforestation, desertification, swamping, etc.). Pollution from working plants can be continuous or casual, and the wastes can have a complex composition. Our modeling is restricted to a single pollutant measured in some units, for example, in percent concentration in air, water, or soil. At a given pollution rate, this concentration may depend on the capacity of the medium to decompose and/or remove polluting materials. Of course, pollutants can pass to another medium, e.g., from air to soil or back, but we look into the pollution dynamics within a single medium. Like other complex systems with numerous interacting parameters, this dynamics cannot be described by a straightforward formalism, but phase portraits capable of predicting the further course of processes can be applied instead of thousands of equations.

The time-dependent concentration of a pollutant is similar to the dynamics of an insect population (see Chapter 1), with the only difference that the future concentration depends on ongoing pollution as well as on the previous concentration. Thus, keeping to the analogy with insects, we assume that the evolution of a native population is disturbed by continuous or casual invasion of insects from outside.

First we leave aside the polluter and study the polluted medium in some selected site with its own properties located at some distance from the polluter. Suppose some pollutant got anyhow into the medium. The time-dependent behavior of pollution can be predicted from the respective phase portrait. We investigate the pollution dynamics over equal time intervals called "years" for simplicity, though, unlike the case referred to in Chapter 1 where generations change every year, industrial pollutants can change their concentration at different characteristic periods depending on their input and removal. The specific length of these intervals is obviously different for different pollutants.

According to the phase portraits method, the concentration of a pollutant at a given point (K) is measured on some day of a current year, say, on December 31; and the measurement repeats in a year and the concentration becomes (M), assuming no additional pollutant input during the year of observation. Then the pair (K, M) is called a "standard observation" of pollutant removal. The cloud of points resulting from a series of observations is plotted in the plane (K, M) to give a phase portrait. There are reasons to assume that M depends mainly on K, though, of course, the removal process can be influenced by casual effects, such as weather, groundwater circulation, etc. Neglecting these fluctuations, suppose that M is a definite function of K: $M = f(K)$, which we call the removal function.

This function, hard to express in equations, is found from a number of repeated measurements (standard observations). Unlike the functions used in Chapter 1, M is always less than K in our case, i.e., the pollutant can only decrease in concentration being altered somehow into other substances presumed to be inoffensive or transferred to other media (from soil to air or water, etc.). We also assume that the pollutant does not reproduce itself unlike the case of a bacterial pollution, i.e., we deal only with nonliving pollutants.

Note that we focus on a single medium and neglect the damage the pollutant might cause when moved to another medium, with a different phase portrait.

A general idea of the removal functions can be inferred from the available experimental data. Assume that the pollutant is removed by either living or nonliving agents. Figure 1 shows a simplified pattern of their combined effect. Inasmuch as, according to our assumptions, $M < K$, the phase curve entirely lies below the bisector. (The analogy with insects would mean that the population becomes extinct without input from outside).

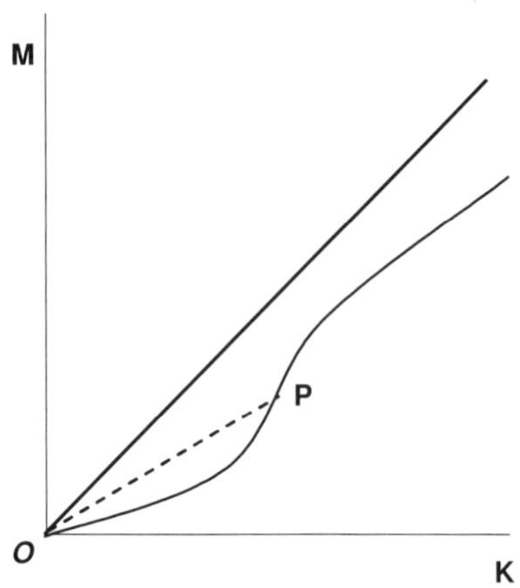

Fig. 1. Typical phase portrait for pollutant removal under the combined effect of living and nonliving agents in a given territory.

At $K = 0$, pollution is zero in the end of a year if the initial pollution is zero, and $M = 0$; therefore, $M(0) = 0$. We assume that for small K the effect from both living and nonliving destructive agents is combined and linear, i.e., the pollutant concentration decreases by a constant factor $c_1 < 1$, that is, $M = c_1K$. Figure 1 shows the corresponding rectilinear segment of the phase curve where the ratio M/K remains constant and below 1, i.e., the segment is at an angle <45° to the K axis. Then we assume that at a higher concentration *K*, the living agent becomes extinct being suppressed by chemical changes but the nonliving agent acts linearly as before; hence, a lesser portion of the pollutant is removed. This means that the ratio *M*/*K* grows at further increase of *K*. Geometrically, the chord *OP* to the point *P* with the coordinates (*K, M*) makes an increasing angle with the *K* axis (Fig. 1). Eventually, the living agent falls out and the nonliving one remains to act still linearly but is less efficient than before: $M = c_2K$, where $c_2 > c_1$. See the rectilinear segment to the right in Fig. 1 where the ratio *M*/*K* is constant, and its extension likewise passes through the origin of coordinates.

There is a less common case when the medium launches some additional mechanisms against increasing pollution. Then, the ratio *M*/*K* decreases after a certain level of pollution, i.e., more pollutant is removed when it reaches a high concentration than when its concentration was lower. The slope of the chord *OP* decreases, and the curve sags down (Fig. 2). However, on further pollution growth the ratio *M*/*K* starts to grow again, and the scenario follows that of the previous figure.

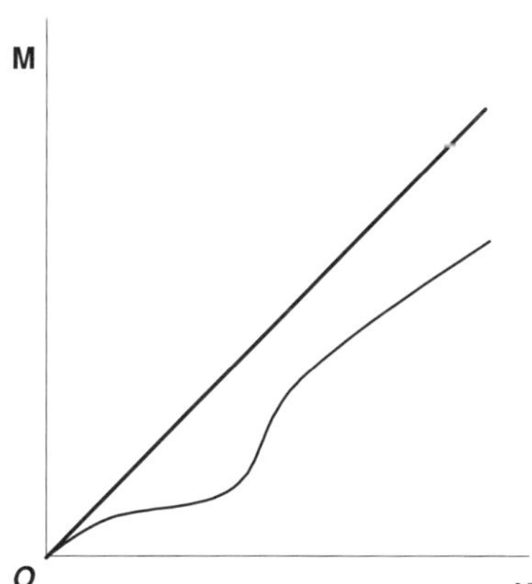

Fig. 2. Pollutant removal in the case when the medium launches some additional mechanisms against increasing pollution.

It is seen from Fig. 1 (or 2) that the ratio is always much lower than 1, that is, an essential concentration change (from *K* to *M*) occurs during a conventional year between measurements. In physics "characteristic time of change" is the time in which some parameter changes significantly but not too much; or in our case

$$a < M/K < b,$$

where a and b are of the order of 1 ($0.1 < a/b < 10$). In Fig. 1 the characteristic time of concentration change is of the order of a conventional year. Concentration decreases little if the observation period is much shorter than the characteristic time, i.e., M/K is about 1. This phase curve nearly coincides with the bisector and has no practical application. Likewise, if observations repeat at a period much longer than the characteristic time, M/K approaches zero, the phase curve coincides with the K axis and is neither useful. Thus, pollution prediction should base on time intervals comparable with the characteristic time of pollutant removal. It can be several decades or several days long depending on the character of pollution and removal, though it is conventionally called a year.

The behavior of pollutant removal imaged in a phase portrait is a good indicator of the medium properties and its response to specific pollution, which remains very little explored. Below we demonstrate its application for predicting the consequences of industrial pollution.

We begin with a single discharge not followed by further pollution and thus can apply the phase portraits of Figs. 1, 2. Single discharges are not typical of continuously running plants but are rather catastrophic events like the Hiroshima nuclear explosion or the Chernobyl accident. Those tragic events stimulated detailed studies of pollution dynamics in different media which made it possible obtaining phase portraits of removal for some chemicals, especially radioactive ones. Thus the catastrophes provided data for science against the ethical law prohibiting experiments on people!

Let K_0 be the pollutant concentration immediately after the discharge and K_1 its concentration in a year, in the absence of further pollution. K_1 can be found from the phase curve, the concentration in two years K_2 can be found in the same way, etc. Using reflection from bisector (see Chapter 1), we easily find that the concentration tends to zero with time, irrespectively of the initial discharge amount (this follows from successive plotting as in Figs. 8 and 9 in Chapter 1). Of course, it is desirable to know the time required for complete pollutant removal which is likewise predictable from the phase curve. It can be long if the initial pollution is very large, i.e., the point P is far to the right (see Fig. 1 or 2).

There is another important note. We assumed before that pollutant removal depends only on its initial concentration but not on how it accumulated. However, long-lasting previous effect from pollution can change the medium properties so that it becomes a different medium at different moments of time, but we assume that it remains the same. Note that unlike the phase portraits below, the phase portrait for a single discharge depends only on the chosen observation place rather than on the location of the polluter and represents the response of the medium to pollution in this specific site.

Single discharges are of course rare being mostly a matter of emergency. Normally running plants cause periodic or continuous pollution for a long time. However, phase portraits for both periodic and continuous pollution (and actually for the general process of continuous pollution) can be inferred from the phase portrait of a single discharge. This important result allows an insight into the mechanism of pollution by a running plant.[a]

First consider the case of a plant that discharges equal amounts of a pollutant periodically at equal time intervals, suppose, every year (the conventional year as we mean it) at 00:00 on January 1.[b]

Let the phase function corresponding to periodic discharges be $g(x)$. It turns out that we can find the function $g(x)$ from the known phase function $f(x)$ of a single discharge. Indeed, let the pollutant concentration on December 31 in a current year immediately before midnight be x and

---

[a] The theorem proved below belongs to V.A. Okhonin, and is published in English for the first time in this book.

[b] Proceeding from the above, the periodicity should be of the same order as the characteristic time of removal. If it is much longer, in the absence of pollution between discharges, the problem reduces to the basic phase portrait for removal; if it is much shorter, we can interpret discharges as continuous pollution and use the uplifted curve of Fig. 6.

the concentration produced by a discharge at 00:00 on January 1 and measured immediately after (prior to the onset of removal) be d_0. This is basic information on the polluting effects of the polluter, along with the yearly periodicity of discharges. Thus, the total pollutant concentration immediately after the discharge is $x+d_0$. This amount is removed during the following year (year of observation), without further pollution, till midnight of December 31 when it becomes $f(x+d_0)$, by definition of the phase function of a single discharge. On the other hand, the phase function of periodic discharges is $g(x)$; repeated pollution (which reduced to the single discharge of January 1 during the year of observation) brings the concentration x to $g(x)$. Thus, on December 31, before midnight, we have

$$g(x) = f(x+d_0).$$

The curve $g(x)$ is obtained from $f(x)$ by a simple shift for d_0 (Fig. 3): this means that the value of g at the point x equals the value of f at the point $x+d_0$ shifted to the right for d_0. Then, the plot of the function g is obtained from that of the function f by shifting the latter to the left for d_0.

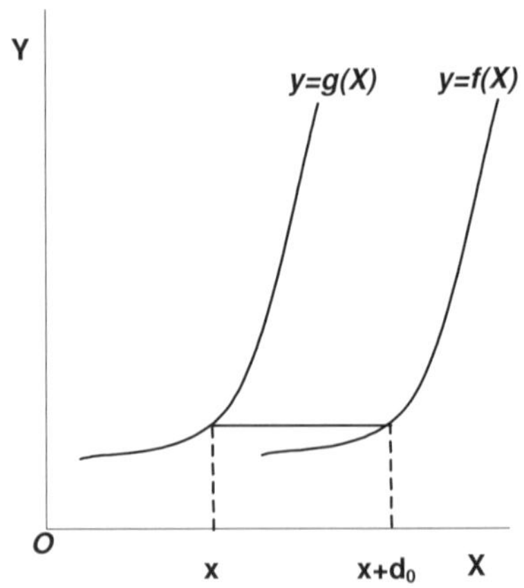

Fig. 3. The phase function of repeated periodic pollution (g) obtained from the phase function of single discharge (f). d_0 is the pollutant concentration immediately after the discharge before the onset of removal.

Thus, the following theorem has been proved:

The phase function of periodic pollution is given by

$$g(K) = f(K+d_0)\ ,$$

where $f(K)$ is the phase function of removal in the medium and d_0 is the concentration immediately after a single discharge.

Shifting the curve $f(K)$ to the left for d_0 gives the curve $g(K)$. The values of $g(K)$ for negative K are rejected as senseless (because the initial concentration is certainly positive). The form of the shifted curve is investigated below.

To apply the theorem just proved, one has to know the phase portrait $M = f(K)$ for a single discharge which can be drawn if the pollutant amount is sufficient to provide a high initial concentration K for such curves as in Fig. 1. As noted above, the initial concentrations are as a rule associated with fatal accidents the consequences of which were studied. Thus the disasters that give no credit to human reason provide information on environmental damage from "normally" running plants.[c]

The translation of the preceding theorem can be applied to the curve $M = f(K)$ of Fig. 1 to obtain the left curve of Fig. 4 (valid at positive K). The curve has a stable equilibrium point 1 at its intersection with the bisector (see Chapter 1).

Now consider the case of uniform continuous pollution. In this case, besides the phase portrait of Fig. 1, one has to know the pollutant concentration in the very end of the first year of plant work (d_1). It can be approximately identified with the "mean annual pollution", i.e., with the annual concentration arising immediately after a short period of operation. This is obviously not exact as pollutants can become partly removed by the time of measurements. Nevertheless we call the concentration d_1 "mean annual pollution".

---

[c] Note that the general case of variable pollution amount can be reduced to a single discharge using essentially the method just described, that is, by summation of concentrations left after the preceding discharges had been removed. This is done by the device of reducing continuous processes to discrete ones well known in mathematical physics.

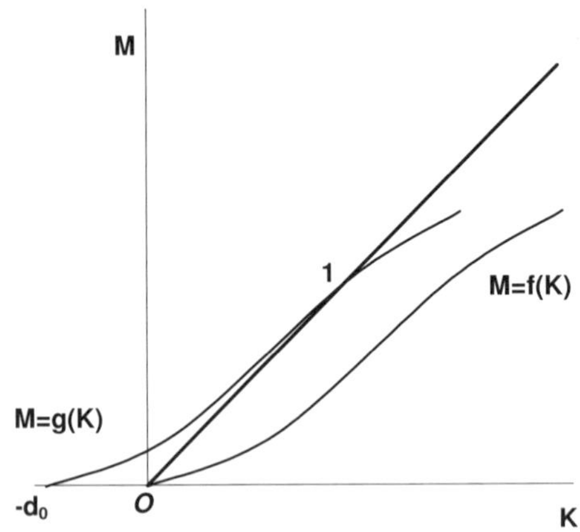

Fig. 4. Phase portrait of pollutant removal at periodic pollution in a given territory (obtained from phase function of pollutant removal at a single discharge). d_0 is the pollutant concentration immediately after the discharge before the onset of removal.

Let x be the concentration at the beginning of the year, remained from the preceding plant activity. In the absence of pollution during the following observation year, the pollutant reaches the concentration $f(x)$ in the end of the current year, by the definition of the phase function of a single discharge. However, this amount should be added with the concentration d_1 produced by continuous operation of the plant during the year, so that the concentration in the end of the year becomes $f(x) + d_1$. According to our definition of $g(x)$, this is exactly the value of the phase function of continuous pollution $g(x)$, i.e., the concentration remaining in the end of the year if it was x in the beginning. Thus, we proved that the phase function of continuous pollution is given by

$$g(K) = f(K) + d_1,$$

where $f(K)$ is the phase function of removal for the given medium, and d_1 is the mean annual pollution.

Each value of g exceeds the corresponding value of f for one and the same amount d_1, which means that the curve moves up for d_1 (Fig. 5).

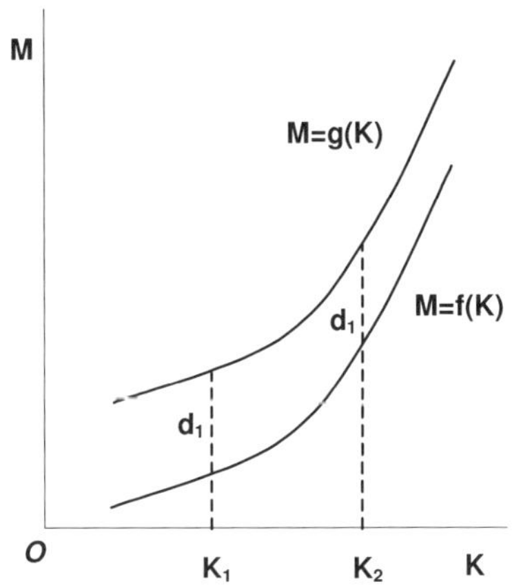

Fig. 5. Phase portrait of pollutant removal at continuous pollution in a given territory (inferred from the pollutant removal at a single discharge). d_1 is the mean annual pollution.

If the removal phase curve for a single discharge has the same form as in Fig. 1 (which is corroborated by reliable data), shifting the curve up for d_1 gives the phase curve of continuous pollution (see below).

We showed that the phase portrait of pollutant concentration for a continuously running plant was obtained from the single discharge curve either by a leftward shift for d_0 (periodic pollution) or by an upward shift for d_1 (continuous pollution). The qualitative results are similar for both cases. We consider the latter case as an example, and the former one can be treated in the same way.

When the curve $M = f(K)$ moves up for d_1, its left end raises from the origin to the point $(0, d_1)$ to arrive at the position above the bisector. On the other hand, at high K the curve $M = f(K)$ coincides with the straight line $M = c_2K$, at $0 < c_2 < 1$. Thus, the line slopes at less than 45° relative to the K axis and is below the bisector at high K, as well as the phase curve. At intermediate values of K several cases are possible.

(1) The curve $M = f(K) + d_1$ crosses the bisector at single point 1 (Fig. 6) and then remains below the bisector. For simple geometric reasons (cf. Fig. 1), this is true if d_1 is not high enough for the points of $M = f(K)$, located far from the bisector as the rise begins, to reach the bisector (see the upper curve in Fig. 6).

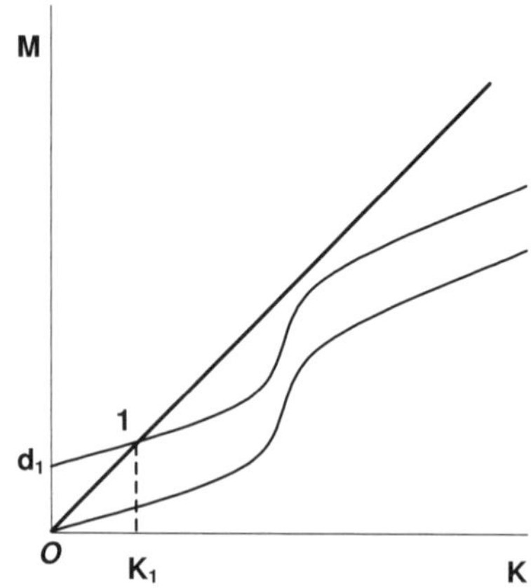

Fig. 6. Phase portrait of pollution from a continuously running plant obtained from that for a single discharge by an upward shift for d_1 at moderate d_1. The system has a single stable equilibrium point (1) and pollution is moderate. With this phase portrait, pollution at point 1 should be kept below maximum permissible concentration (MPC).

Reflection from bisector (Chapter 1) shows that this curve has a single point of stable equilibrium (point 1) with its K-coordinate K_1. Then for $K < K_1$ the K-coordinate of the point moves with time to the right and arbitrarily approaches point 1 in several steps, each corresponding to a conventional year. At $K > K_1$, the point of the phase curve moves to the left likewise toward point 1. Thus, point 1 represents a state with a stable concentration of pollution K_1. There are no other equilibrium points as the phase curve crosses the bisector nowhere else. The concentration K_1 depends on the mean annual pollution d_1. The phase curve allows us to predict the stable concentration K_1 and thus to decide whether the polluter is tolerable; otherwise the project of constructing such a plant should be refused or an operating plant should be closed out.

(2) The curve $M = f(K) + d_1$ crosses the bisector at three points (1, 2, 3) at larger d_1: as d_1 grows, the curve $M = f(K) + d_1$ rises so that its convex part touches the bisector and then rises above it at some $d_1 = d_{1a}$ (see the upper curve in Fig. 7).

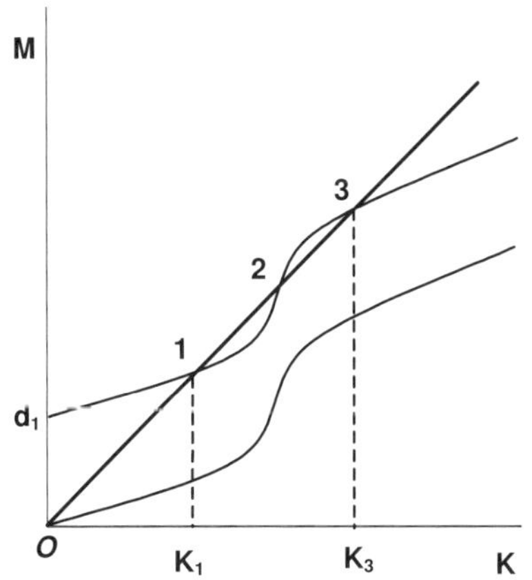

Fig. 7. Phase portrait of pollution from a continuously running plant obtained from that for a single discharge by an upward shift for d_1 at d_1 greater than in Fig. 6. The system has three equilibrium points (two stable and one unstable). Stable point 1 corresponds to relatively low pollution (comparable to MPC) and stable point 3 corresponds to high pollution, usually many times the MPC.

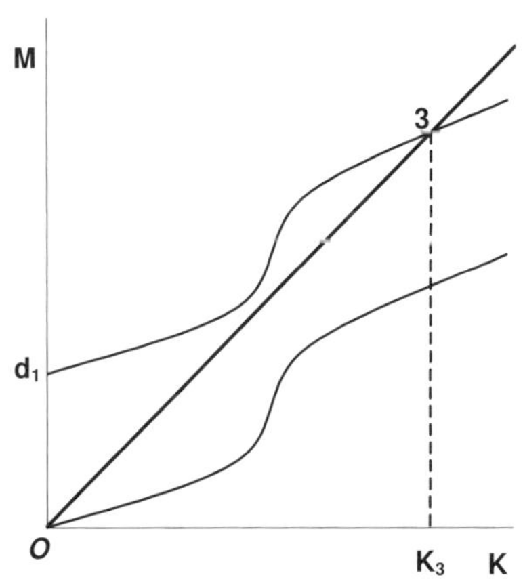

Fig. 8. Phase portrait of pollution from a continuously running plant obtained from that for a single discharge by an upward shift for d_1 at d_1 greater than in Fig. 7. The system has a single stable equilibrium point (3) corresponding to a very high pollution and the environment collapse of the territory.

We call d_{1a} the first critical value. At large K the curve is parallel to the straight line $M = c_2K$ (which is at < 45° to the bisector) and eventually sags below the bisector. Thus the curve $M = g(K)$ does cross the bisector at three points (1, 2, 3).

(3) At greater d_1, the curve $M = f(K) + d_1$ again crosses the bisector at a single point (point 3), whereas point 1 disappears (Fig. 8). Indeed, further growth of d_1 leads to $d_1 = d_{1b}$ (the second critical value) when the concave part of the curve touches the bisector and then rises so that both points 1 and 2 disappear. Point 3 however remains, as the curve again sags below the bisector at greater K. The pollution concentration corresponding to K_3 of point 3 is in this case even higher than in case (2). For most pollutants this concentration level is fatal for environment, which should be taken into account to plan a new safe technology.

The point of stable equilibrium (point 1) is of greatest practical value as it represents the operation of all normal (environment-safe) plants. We are to find the corresponding pollutant concentration (K_1).

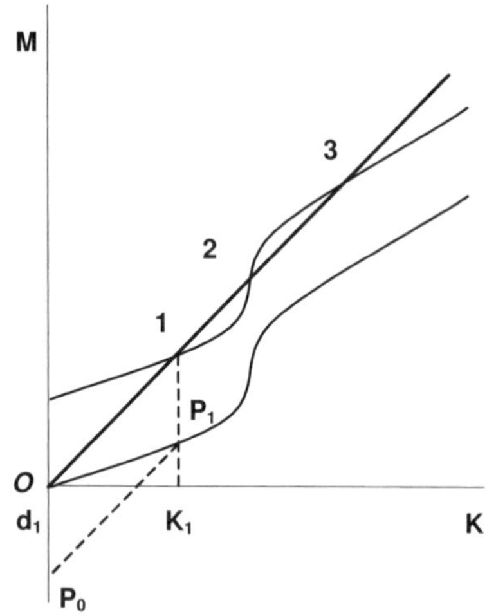

Fig. 9. Pollutant concentration at point 1 of stable equilibrium (K_1) corresponding to the work of "normal" plants, obtained graphically from the basic phase portrait for pollutant removal.

As all our curves are empirical, the required K_1 is obtained graphically (see Fig. 9). The lower curve in this figure is the phase portrait of removal $M = f(K)$, the upper curve is the phase portrait of continuous pollution $M = g(K)$, obtained from the former by rising for d_1. If we put the segment OP_0 of the length d_1 along the M axis downward from the origin and then draw a straight line through P_0 parallel to the bisector until it crosses the lower curve at P_1, the vertical through P_1 crosses the bisector at a point shifted above P_1 for d_1 and thus belongs to the upper curve. The intersection of the upper curve with the bisector is nothing but the sought equilibrium point 1 (Fig. 6). Thus, the K-coordinate of P_1 denoted as K_1 is equal to the M-coordinate of point 1 and the latter consists of the segment K_1P_1 of the length $f(K_1)$ and the segment $P_1 1$ of the length P_0O, i.e., $K_1 = f(K_1) + d_1$ and K_1 is indeed the sought root of the equation $K = f(K) + d_1$.

Environmental disaster

The pollutant concentration is always considered at a certain site around the polluter. In the simplest case of a homogeneous environment its response to pollution is the same everywhere, i.e., the removal phase curve is the same for all points reached by pollution: $M = f(K)$. Remind that this curve records the removal of the initial concentration K, whichever be its origin, and depends only on the properties of the environment, assumed to be homogeneous.

The mean annual pollution d_1 is, by definition, the concentration produced by the polluter for a year, measured in the very end of the year, under the assumption that the plant had not worked before. The result of course depends on the place of observation; d_1 must decrease away from the polluter, as pollution spreads over a larger area. The state of environment at some point P of the territory (Fig. 10) largely depends on its distance from the polluter located at the point O. Neglecting the wind rose (prevalent direction of air flows) we assume that d_1 depends only on the distance OP, and is a decreasing function of the latter:

$$d_1 = S(OP).$$

This function is constant at equal distances *OP*; thus, it takes a constant value on every circle centered on *O*, and the phase function of continuous pollution $g(K) = f(K) + d_1$ is constant as well.

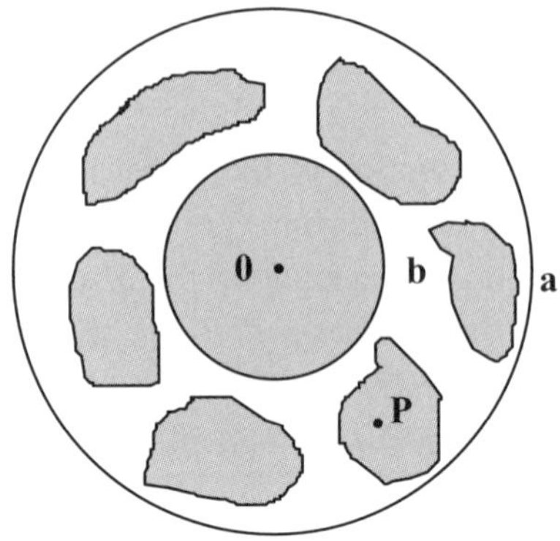

Fig. 10. Environment in a territory with a polluter located at the point *O*. Wind rose neglected. See text for explanation.

Now consider the following cases.

A. In the immediate proximity of the polluter *O*, $d_1 < d_{1a}$ where d_{1a} is the first critical value (see above). Then, d_1 being a decreasing function of distance, the inequality fulfills universally. Hence, case (1) of our analysis, with a single stable point (point 1), occurs everywhere. The concentration K_1 corresponding to the stable equilibrium decreases when the observation place is moving off *O*, and this concentration at different places has to be checked against the assumed permissibility criteria, at least the maximum permissible concentration or MPC (for more details of MPC see Chapter 8).

B. In the immediate proximity of *O*, $d_{1a} < d_1 < d_{1b}$ where d_{1b} is the second critical value. Then there is a circle (*a*) with the center *O* along which $d_1 = d_{1a}$, $d_1 < d_{1a}$ outside it, and $d_{1a} < d_1 < d_{1b}$ within it. Outside the circle *a* only case (1) is possible. Case (2) works within the circle, with two possible stable levels of pollution (1 or 3) actualized casually. As the environment is imperfectly homogeneous anyway, some areas within *a* are polluted at the concentration K_1 and others at K_3 (the latter are gray areas in Fig. 10). The pollutant concentration in these places often exceeds the limit resistance of vegetation, which becomes evidently depressed in more vulnerable localities. Such a patchy pattern of the territory around the polluter is indicator of type 3 steady pollution, i.e., of environmental disaster.

C. In the immediate proximity of O $d_1 > d_{1b}$. Then there is a circle (b) along which $d_1 = d_{1b}$ (Fig. 10). Thus, only steady pollution of type 3 is possible within b, usually with a very high concentration K_3 (the phase curve can cross the bisector in Fig. 8 arbitrarily far). Within the shaded circle of Fig. 10 vegetation is fully depressed, and people working at the plant or living in its vicinity should be aware of the risks they run.

The patched pattern of the territory between a and b is a case observed quite often. However, the absence of evident pollution outside a does not mean that the place is good for living. The situation rather requires further studies.

We assumed above that there were no prevalent wind directions in the territory, but if there are any, the picture of disaster (Fig. 10) deforms, remaining qualitatively the same. For example, the circles a and b give way to elongate ovals aligned with a prevalent wind direction.

Note in conclusion that the dynamics of environmental damage remains underexplored, and very little is known about natural processes of pollution removal. The available data are restricted to quite numerous measurements which, however, need systematizing and interpreting. For this the phase portrait approach is an efficient tool. It can be applied to obtain the removal phase portraits for different specific media and pollutants on the basis of the existing and new field data. The phase portrait predictions are useful for both mitigation of current damage and assessment of future environmental risks.

Chapter 8

Fining and Environment

Experience shows that fining is an efficient economic measure against impudence of polluters, along with administrative shutdown of enterprises. Of course, the fining works only in societies that live by law and order. Otherwise, the fines would remain on paper or become a pretext for racket.

At present, pollution control and the related fining policy are based on the standards of maximum permissible concentration (MPC) and maximum permissible discharge (MPD) of pollutants.

The concentration of a given pollutant in a given medium is measured (or supposed to be measured) in mass units per surface area (on land) or in volume units (in water or air). The measurements are often occasional in space and time and often follow official regulations rather than scientific knowledge. There are legally specified maximum permissible concentrations (MPC) for different pollutants in different media above which pollution is assumed harmful for humans and domestic animals, and thus punishable. Of course, MPC estimates depend on the current state of medical and biological knowledge and are at least arbitrary. Therefore, it is safer when the MPC levels are underestimated.

Unfortunately, the true meaning of "maximum permissible concentration" is often unclear even for educated public. Many people believe that the existing MPC standards would be set in a way to prevent any damage to human health and that pollution below MPC would be inoffensive. Therefore, it is supposed that keeping to the MPC restrictions ensures avoiding any environmental damage. Public opinion and mass media take heed only of surpassing MPC and calm down as soon as pollution reaches a lower level. This attitude stems from misinterpreting the meaning of MPC.

To assess the dismal reality behind this abbreviation, one has to bear in mind that harmful discharges from plants, vehicles, etc. cause a proven statistically measurable increase in certain diseases and the related death rate. For example, many pollutants may cause cancer, or, as people can suffer from cancer without each specific pollution, at least increase the risk. One can estimate the number M of cancer cases in a locality with a population N. Then the M/N ratio is the cancer risk for this population. In Fig. 1 this probability f is plotted against the pollutant concentration U. The same relationship is typical of many other diseases and many kinds of pollution.

The pollution-dependent death rate curve shows similar behavior. Of course, people fall sick without pollution as well. For example, cancer cases often have no evident cause and may appear accidental. Yet, accidental events are amenable to statistical account, and the risks with and without pollution can be compared. The statistics shows the percentage of population that can fall sick though cannot account for the life of each individual.

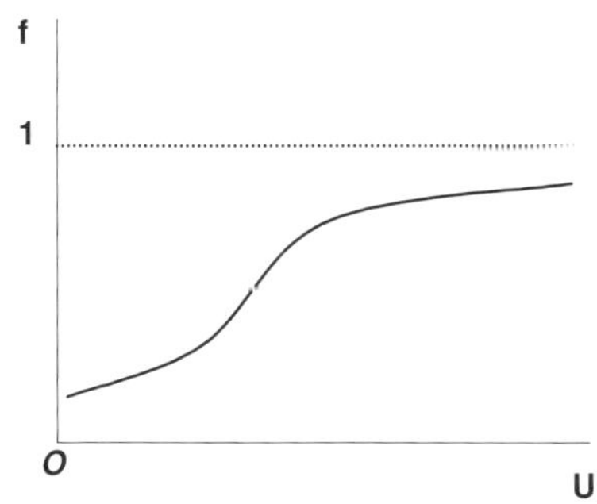

Fig. 1. Health risk f as a function of pollutant concentration U for a locality.

The risk $f = 0.001$ at $N = 100{,}000$ indicates that 100 people can fall sick in the locality. Some change in life conditions, say, putting in operation a new plant, can increase the risk to $f = 0.0012$, i.e., twenty more people will fall sick. Statistical estimates are never exact in the sense that 118 or 121 and not 120 people may suffer and nobody knows their names but these estimates are corroborated by experience. If the risk f is known from experience, one can estimate the morbidity increment related to operation of another new plant with the same pollution rate.

The Z-shaped curve in Fig. 1 is typical of many biological processes. The risk f is nonzero (though low) already at zero U and grows with the latter. The growth is first slow and is getting ever faster to approach unity at high U, i.e., nobody can escape falling sick. This is, for instance, the case of workers at cement plants or mines who inevitably suffer from silicosis.

MPC is set at the level where the curve $f(U)$ is rising quite slowly, though the specific meaning of what is *slowly* often has different interpretations. This arbitrary decision making commands the number of sick people above the natural norm. MPC standards in some countries are so "generous" that the death rate increase reaches 10-15%, i.e., ten to fifteen people more per every hundred die from a certain disease as a result of some specific pollution. This is actually a deliberate sacrifice to economic interests of industry or car use. The opponents of environmental restrictions usually claim that neglect of these interests and conveniences would decrease the life standard of the population as a whole. Debating MPC standards in parliaments is in fact a trade off to decide how many lives to sacrifice to the Moloch of mass consumption. The society would hardly accept the sacrifice if the names of people condemned by this or that regulation were known: Our western civilization, as it is now, shuns the idea of sacrificing few for the common good of many. But anonymous lives are hypocritically let to be sacrificed.

MPC standards are based on the following procedure. The exact behavior of the curve $f(U)$ is, of course, unknown. It is obtained somewhat arbitrarily by drawing a smooth line through the points that represent the available data, though the latter can be inaccurate and most often incomplete. Errors are estimated by mathematical statistics where the uncertainty of the resulting curve is measured by a small positive number ε. It means that the true (and unknown) curve $f(U)$ is to a large probability constrained between $f(U) - \varepsilon$ and $f(U) + \varepsilon$ (Fig. 2). The difference between the true and hypothetical curves is assumed to be within ε.

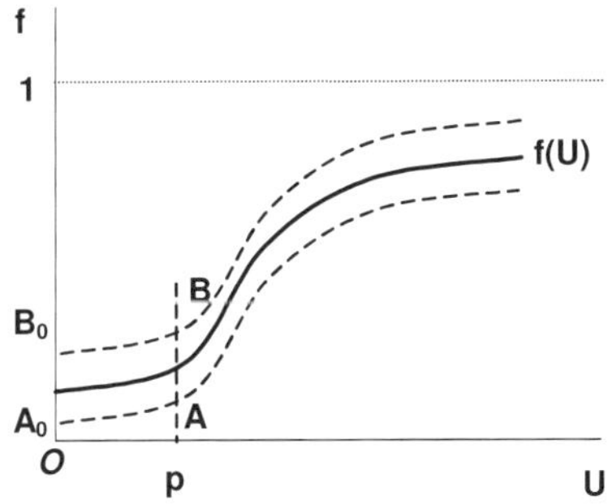

Fig. 2. Estimating MPC. Solid line shows the hypothetical $f(U)$ curve. Dashed lines constrain the error band.

Now we select some MPC value p and draw a vertical line upward from p to cross the error band along the segment AB. On the other hand, the f axis crosses the error band along the segment A_0B_0. The true risk curve $f(U)$ crosses the two segments at the points C and C_0, respectively. See this part of the picture enlarged in Fig. 3.

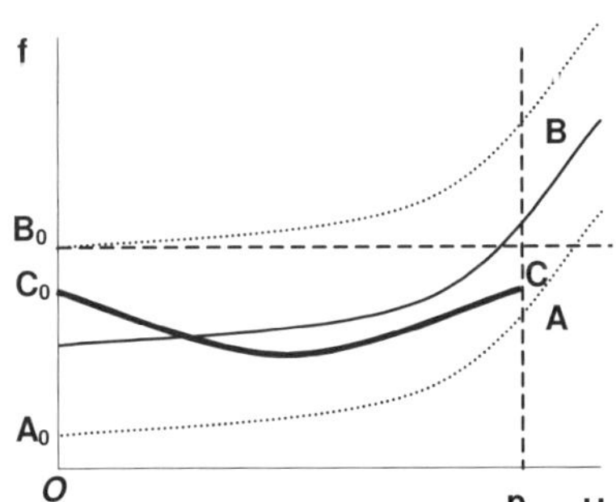

Fig. 3. Estimating MPC. Enlarged from Fig. 2. Solid line shows the hypothetical $f(U)$ curve. Dotted lines constrain the error band. Heavy line shows the true $f(U)$ curve.

By definition of the true curve, the f-coordinate of the point C or $f(p)$ corresponds to the risk at the pollution p and C_0 or $f(0)$ to the risk in the absence of pollution. If the two probabilities are equal, it means geometrically that the points C and C_0 are at the same height which may occur if p is not too high. However, if the risk in the presence of pollution p can be equal to that in the absence of pollution (i.e., it does not contradict the available data accuracy), it is impossible to rigorously prove that pollution is responsible for increasing the risk. This "unprovability" suggests the following procedure for estimating MPC. A horizontal line is drawn through B_0 to cross the lower limit of the error band and the U-coordinate of the point A slightly below this intersection is taken for MPC. It is easy to check that at this p the segments AB and A_0B_0 are crossed by the same horizontal line.

Therefore, the ordinary statistical procedures are hypocritically used to select MPC such that it allowed the responsible people to deny the damage from pollution. Thus the impossibility to prove the danger is passed off as a proof of safety! This choice is essentially based on the old saying that one is innocent until proven guilty, and all doubts are interpreted in favor of business rather than of human lives. Improvement of statistical methods can reduce errors and narrow down the error band to lower the obtained MPC standard. However, there are many ways to keep arguing that data are still unreliable and the lower standard has no solid ground. Of course, decisions depend in this case on the honesty of environmental and medical people.

Certainly, limitations on pollutant concentration are useful but insufficient as this restriction leaves the polluter with the chance to disperse its wastes over a larger area which reduces the concentration but pollutes a larger territory. That is one reason why factory chimneys are made so high. The other pollution parameter, maximum permissible discharge (MPD) which measures the amount of a discharged pollutant, aims at preventing this practice.

Restrictions based on the legal measures, insufficient as they may be, are useful, and their violation reasonably outrages people. For example, the concentration of heavy metals in soil often reaches hundreds and thousands MPC in urban industrial zones. These facts are broadly discussed in mass media but the cases of mere violation of the existing laws are beyond the scope of our book.

Yet even the MPC and MPD standards taken together cannot ensure the appropriate pollution control. It requires developing a system of parameters as a basis for phase portrait modeling that can predict the pollution dynamics in specific localities. Those who are guilty of environmental damage have to be punished, and fining is a workable way to mitigate the crisis, unlike the tricks of propaganda.

Of course, any discussion of fining appears pointless unless the subject of fining is clear, which rises the question of the appropriate estimating of the environmental damage. Nevertheless, the discussion is timely. The effect of fining regulations is similar to any other economic motive (see Chapters 9-14). Without going far into details of business thinking, it is quite easy to predict how people would respond to any money loss. Suppose the environmental damage is measured against some parameters, even against the imperfect MPC and MPD standards. If there are any reasons to believe that their lowering will be good for environment, it is useless to wait until science develops better standards and, more so, until these will be accepted by the authorities. Fining based on the existing standards can work if it really makes polluters take measures to reduce the damage. As new standards appear, the previous ones have only to be shifted and the new fining system will work in the same way because polluters will follow the same market motives. Environmental measures may include, for example, installing pollution control facilities or even rebuilding the technology.

The theory of fines outlined below is indifferent to particular offences of polluters amenable to fining. What is important is that elimination of the offences mitigated environmental risks and that the fining system made the offenders — which are guided uniquely by their private interest — stop polluting the environment.

So far we treated only environmental damage and pollution as its simplest kind. In this respect, there are two parts concerned: population and polluters. Speaking about fining implies the existence of another part, that of fining agencies established by authorities. It is naive to imagine that fining agencies would be guided uniquely by public interests. Few enthusiasts do exist but people in the average, which is the subject of modeling, most often pursue their own ends. Therefore, we proceed from the assumption that fining agencies, remaining within the established regulations, would apply them at their will. Our model neglects the situation when fining agencies break the law being employed by polluters. Fining works only if direct law breaking never remains unpunished. We show further how fining regulations can be evaded and how the abuses can be controlled.

Figure 4 presents the simplest case of the environmental effect of fining.

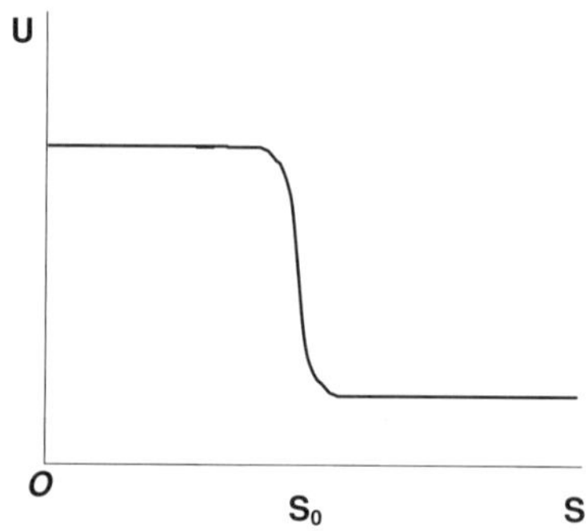

Fig. 4. Environmental effect of fining. S is the size of fine per pollution unit; U is environmental damage (measured as pollutant concentration or otherwise).

The size (S) of fine per pollution unit (the latter can be measured in different ways, as mentioned above) is established by authorities and can change with time (its time dependence is not shown in Fig. 4 which gives rather environmental damage as a function of fine size).

The polluter shows no response to fining till some threshold value S_0, i.e., it prefers paying to installing pollution control facilities. (Mind that direct law breaking is not allowed and the polluter really pays the fine if the offence is proved). At some critical S_0, it becomes more profitable to set up wastes control than to continue paying. As a result, pollution diminishes, and the curve passes from the upper stair step to the lower one (Fig. 4). This new level of pollution satisfies the polluter, as the control facilities were installed exactly for that reason. On the other hand, the abrupt fall in pollution is good for environment near the critical point S_0 and thus satisfies the population. Yet, it is the fining agency that is not satisfied!

The usual practice for fining agencies is as follows. The total amount D of payments they raise during a year depends on the fine size S per pollution unit (Fig. 5). The linear segment of the curve shows the proportional growth of D and S, at the constant level of pollution. Fining agencies live on the amount D, their activity being stimulated by leaving them a part of the payments. Of course, local authorities profit from this money as well, but we call it the gain of fining agencies for simplicity.

Fining agencies continuously reap their profits up to the level S_0 while nothing changes in the environment. This is obviously a sheer racket, an exaction imposed whichever be its pretext. The environmental racket never reduces pollution in the same way as an ordinary racket never makes for any decrease in crime.

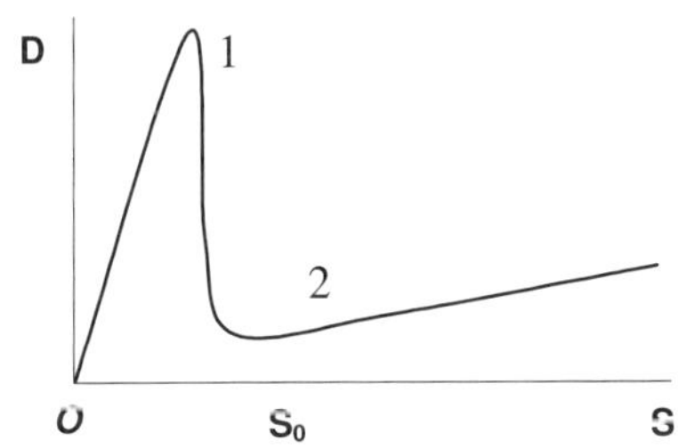

Fig. 5. Yearly amount of payments as a function of fine size. S is the size of fine per pollution unit; D is the total amount of yearly payments got by fining agencies.

The fine size S is legally established (by public or local authorities) at some level but the actual size depends mostly on fining agencies which can increase it by increasing the frequency of payments. The agencies can neglect some offences either raising bribes or just avoiding too rigorous fining which may really make the polluters set up pollution control facilities and thus reduce their gain (see the fall in D after the critical point in Fig. 5). Therefore, fining agencies are quite flexible in applying (or not) the established regulations, almost unconsciously, being guided by their intrinsic main interest. Thus, polluters respond to the actual rather than officially established fining level taking into account the practiced mean fining size. In fact by S we meant this latter size.

Fining agencies obviously do not care about environment but just optimize their gain. They prefer keeping to the level slightly below S_0 (the case of Figs. 4 and 5) gaining the best profit from their racket but preventing polluters from installing pollution control facilities to escape fines. This level corresponds to the highest D (point 1) in Fig. 5.

Suppose, however, that pressure from public opinion outraged by the ongoing pollution makes authorities rise the fining level to S_0. Then pollution can be restricted to the lower level of Fig. 4 at the same established fine size acceptable for both the population and the polluter who brought pollution control to the level S_0 where it was ready to pay. Yet, the situation is undesirable for the fining agency as it looses its gain from racket and, possibly, also for the authorities as they loose their share of payments. The authorities get the greatest part of the raised payments and give up a fixed amount to fining agencies, and thus they loose when the critical level of fining decreases (see the D-coordinate of S_0 in Fig. 5).

Then, propaganda mechanisms are launched to work over the public opinion. People are told that the reduced pollution is still unsafe, the taken measures are insufficient, and greater fining is required to force the polluter to meet the public interest. All this verbiage, of course, finds support with medical men who do care about the effect of pollution on people's health, and with politicians who repeat the medical arguments and invent new ones. The propaganda driven by strive for racket makes the amount S_0, acceptable for public, the point of unstable equilibrium.

This triggers further escalation of fines. As the pollution remains constant at the lower level of Fig. 4, the gain of the fining agency grows linearly, i.e., proportionally to the fine size, as well as the gain of the interested authorities. Thus the polluter pays ever more for the same pollution whereas the environmental risk invariably remains at the level corresponding to the point where the pollution control facilities had been set up. The new level of payments is still racket, though the gain is most often lower than before (lower than the highest point in Fig. 5), as it is impossible to prevent people from unrestrained grabbing profit from their position. The money coming from fining is sometimes spent on some municipal economy improvements but the fined companies usually put the expenses over the consumers rising the prices as high as they can. Nevertheless, the new racket level corresponds to a better environmental situation than that of the first linear segment, which is the reason for the fining system.

The scenario associated with further rising of the fine size is as follows. The second linear segment in Fig. 5 slopes more gently than the first one, and the difference is obviously proportional to pollution decrease. At a too high level of fines, the payments become again unbearable for the polluter and the latter has to close out if cannot find a way to further reduce the pollution (or to transfer the enterprise to a third-world country). Of course, this is against the interests of the fining agency as it cuts off its profits or even threatens its very existence. Therefore, it fixes the fine size slightly below the maximum acceptable for the polluter thus preventing the latter from closing out (Fig. 6). Point 1 in Fig. 6 corresponds to the highest level of "useless" fining, point 2 to the lowest (unstable) level of fining after the onset of pollution control, and point 3 to the highest level of "useful" fining. The fall of the curve after point 3 in Fig. 6 means that further increase in fine size decreases pollution because the polluter falls out. Competent fining agencies would not let the closure as it would mean the cessation of payment.

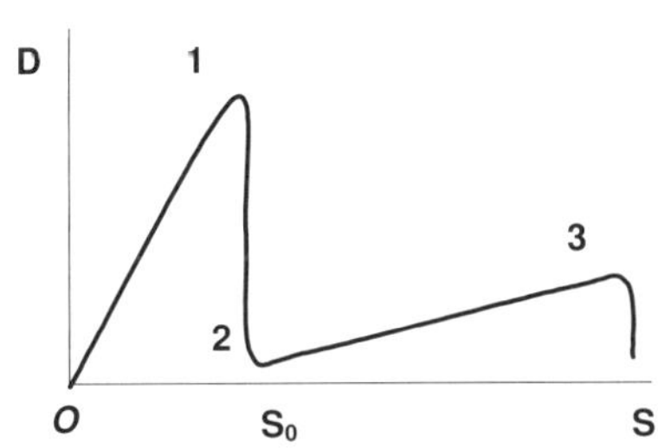

Fig. 6. Yearly amount of payments as a function of fine size. S is the size of fine per pollution unit; D is the total amount of yearly payments got by fining agencies.

A reasonable strategy of the polluter is to install new pollution control facilities which can decrease the level of fining in point 3 and at the same time reduce the environmental damage, though the situation will remain in the stable point 3. Thus, fining works mainly at the stage where pollution control facilities exist but can be further improved. There we pass from Fig. 6 to Fig. 7 showing that improving the existing pollution control is advantageous for the polluter though unfavorable for the fining agency which looses the gain. The situation when the fining agency behaves according to Fig. 6 and imposes high fines while the polluter is flexible enough to improve pollution control (Fig. 7) means that fining really works.

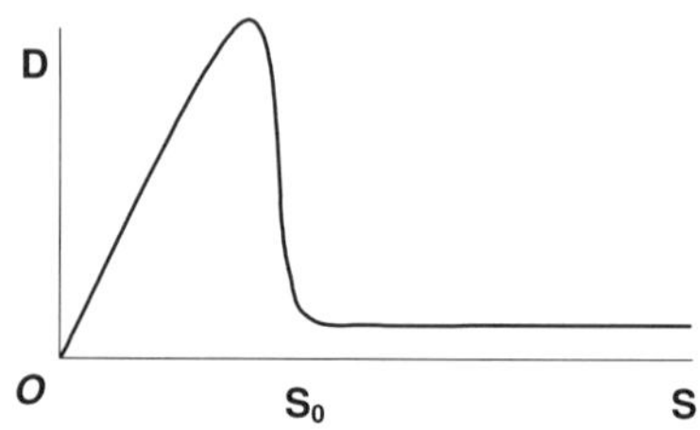

Fig. 7. An efficient fining system. See text for explanation.

The very existence of stable point 3 in Fig. 6 corresponding to building workable pollution control facilities depends on the efficiency of the polluter itself: the more efficient the plant the easier it can afford setting up and improving pollution control facilities without losing in efficiency. Thus points 1 and 3 are spaced farther. If they are far enough, the fine size at point 3 is as high as to allow the fining agency to restrain from the temptation of keeping to the unsafe position of point 1, that of being more indulgent and thus deliberately reducing the fine size.

The choice of the fining agency between strategies 1 or 3 also depends on the efficiency of pollution control. The agency would profit from a not very efficient pollution control and follow strategy 3, and it would prevent the polluter from efficient pollution control and keep to strategy 1. Nowadays strategy 3, or even the pattern of Fig. 7, are often practiced in technologically advanced economies.

Of course, the invention of new pollution control facilities, as well as any technological novelty, not necessarily depends on the willingness of polluters to incur the expenses. Important inventions often appear in small private laboratories run by individual enthusiasts at their own peril rather than by the research and development institutions that belong to large companies. Unlike mere improvements, true inventions are unpredictable. Furthermore, not everything is bought, in environmental control as anywhere. Payment forcing often does not help mitigating the environmental risks.

As mentioned above, the fining agencies can actually change the legally established fine size by varying the frequency of exaction according to their benefit. Thus, the above model included the actual fine size proportional to the exaction frequency. This scenario often runs as if by itself. A fining agency sets up a control service which imposes fines on polluters, however, trying to prevent these from building pollution control facilities. The fining agency thus feeds the municipal budget and ensures a decent salary (as part of the raised payments) for its own employees. This behavior causes no doubt as it appears to be "right" and to agree with the common ways. But it does not help the environment. Blindfold following the common ways is unacceptable as the behavior of the masses (including fining agents) is guided by their intrinsic private profit motivation.

This motivation is recorded in the realistic model of fining below. As soon as fining becomes less paying, fining agencies grow more "indulgent": not all polluters are subject to fining, fewer cases find a true bill, etc. As a result, the polluters become more careless and offences become more frequent. This is a random process often involving many polluters, and fining agencies usually feel the general tendency. Then they can again push up the rigor, and so on. Every following level of control, or the actual fining regulations, is defined by the previous one, as is predicted by the phase portrait in Fig. 8.

The period of changes in the fining policy can be different — every year or every five years, or any — but is felt well by the interested parts. The actual fining policy also depends on the relations between authorities and polluters. This relation is essential even in legal states where local decision making and public opinion are respected as well as the laws.

The phase portrait of Fig. 8 has three equilibrium points. Points 1 and 3 can be stable but are most often unstable in the sense of the quasi-chaotic behavior when the curve crosses the bisector at a high angle (see Chapter 1).

The fine size oscillates at a variable amplitude around these points. Point 2 is unstable in the usual sense, i.e. the process of changes in the

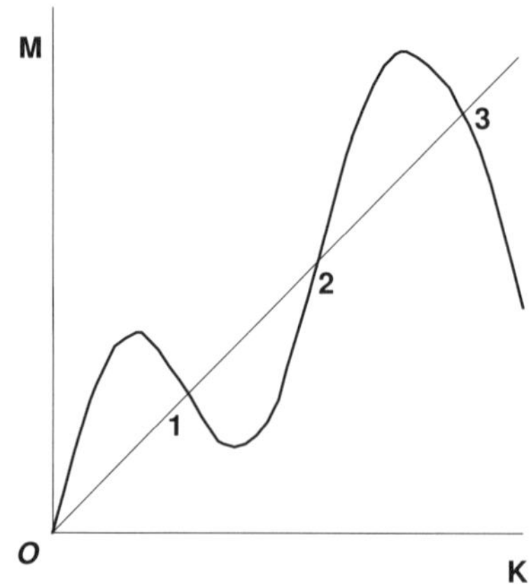

Fig. 8. Dynamics of fine size. *K* shows the actual fine size at some stage of fining policy; *M* shows the respective fine size at the following stage.

fining policy does not stay in its vicinity long enough. The three points correspond to the respective points 1, 2, and 3 in Fig. 6. Points 1 and 3 are especially attractive for authorities and they keep to staying there if have enough power to impose the required level of fines on polluters. Otherwise, fining reduces to the level practicable in the given conditions. Oscillations about the equilibrium points correspond, for example, to a situation when a nuclear power station is closed under pressure from the public and is put into operation again to cover the energy deficit, which provokes new public complaints. Or, we can cite a less tragic example of fining car drivers. Of course, police hardly would be really interested in making the drivers respect the traffic code, and drivers usually know well at which stage the fining policy is at the moment. Controlling surveys often proclaim "combating campaigns" against something, as if anybody were really going to combat anything!

The curve segment between points 2 and 3 is favorable for environment as high fines really push up building and improving pollution control facilities.

Note that environmental measures are often taken between two extremes: the interests of polluters to pay low fines as environmental racket instead of putting up pollution control facilities and the fanatic claims of the Greens to rise the fines as high as to ruin polluters and shut down any industry.

The dynamics of public concerns associated with aggravation of environmental risks in Russia was the subject of an interesting study [Krylov, 1995]. It revealed a well pronounced regularity in the behavior of people (Fig. 9). In Fig. 9 the level of pollution *U* is plotted against public environmental concerns *T*. Estimating public concerns is of course difficult and is rather a matter of applied sociology. However, we take these estimates on trust (unlike all other estimates referred to in this book) and do not discuss how good are their grounds. At a moderate level of environmental damage, the public is especially severely outraged (see the left-hand peak of the curve). This may be due to the onset of a campaign in mass media or to the freshness of the problem. The peak is followed by rapid decrease while the damage continues growing.

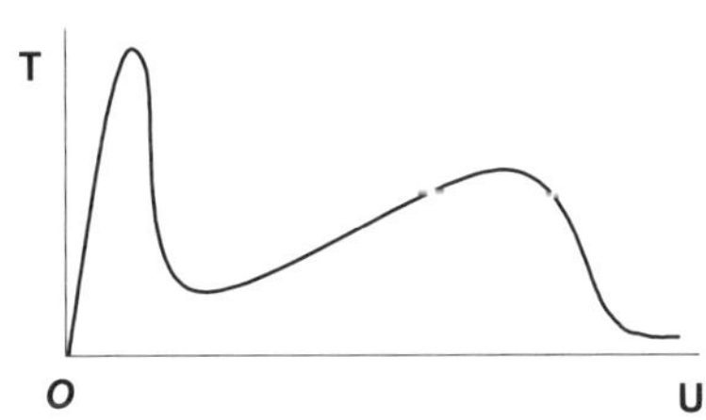

Fig. 9. Behavior of people as a function of environmental damage. *U* is environmental damage (measured as pollutant concentration or otherwise). *T* shows public concerns.

Later on, the consequences of the progressing damage become evident and public concerns show another rise which faithfully reflects the true interests of people. At this stage, more or less successful environmental measures are taken and most prudent families leave the threatened land before the measures prove (or not) to be efficient, as it was after the Chernobyl accident. Finally, at a very high level of damage, the remaining population, usually a great part of the original number, is no longer concerned about environmental risks and resigns to the inevitable evil. This is the common scenario of public behavior, and its final stage is the most dangerous [*Krylov*, 1995].

The today's environmental movements apparently not always strive for overcoming the crisis. The agitation of some Greens, most often the least competent in environmental issues, joins the mediocre propaganda of some extremists, either "right" or "left". Finally, the origin of the Green movement — as any public movement – inspires selfish interests of insolent politicians who abuse any public concern to meet their own ends. They can use the environment-oriented attitudes to build new bureaucratic offices and eventually to strengthen the political controls over the citizens.

Economy and environment

In conclusion we outline possible ways of interaction between environment and economy with an example of some locality that hosts some industry. Let the production of the industry be P (no matter how it is measured) and the amount of pollution per production unit be ε. Then the total amount of pollution is εP. Note that our model includes a single pollutant and its respective share of pollution. There is a maximum total discharge the land can safely rework into harmless materials or assimilate it in a concentration tolerable for people. This amount A is called the carrying capacity of the environment in a locality. Then, the industry is obviously acceptable at $\varepsilon P < A$.

At εP exceeding A, the industry is similar to a parasite that has killed its host. Indeed, people will be unable to live on the polluted land and the polluter should be closed out. Of course, a wise parasite would never bring the matter to that end. For our model locality, the unlimited growth of production P at invariable environmental technologies and the carrying capacity of the territory brings to an environmental-economic collapse, which is the case of an unwise parasite according to our analogy. Fortunately, the technological advance makes it possible to change the amount of production as well as the efficiency of environmental measures and even the total carrying capacity of the territory. Therefore, the three parameters are in fact variable. Pollution can be reduced by improving control technologies and thus becomes a function of time t, i.e., each moment of time has its corresponding $\varepsilon(t)$. The locality capacity can likewise grow and A likewise becomes a function of time $A(t)$. The production P can change if $\varepsilon(t)\,P(t) < A(t)$.

Thus, the three parameters ε, P, and A should change with time in a way to maintain this basic inequality that protects the safety of the environment in the locality. Therefore, the restriction on production at any moment of time is given by $P(t) < A(t)/\varepsilon(t)$.

These straightforward considerations help overcoming the environmental fatalism. Indeed, pollution can be cut down to zero by achieving full recycling of wastes. If this ideal solution is unfeasible, more efforts can be put into increasing the carrying capacity of the environment $A(t)$.

The latter choice finds many instructive examples. Primarily, it is the example of Holland which used to be an unfavorable place to live. The Batavian marshland was no attract even for conquerors. The capacity of the environment was then very low for any production. However, the Dutch were then interested mainly in agriculture and completely changed the soil and water of the country through ages of hard work. Today they keep doing so using the advanced technology. They won an essential part of land from the sea while the capacity of the latter was formerly restricted to fishing or navigation.

Another example is the story of St. Petersburg, the former Russian capital, likewise built in place of uninhabitable marshland. (Pushkin tells it in the wonderful beginning of his *Brazen Rider*.) Peter the Great ordered that office buildings for the future capital were put up together with private houses for the nobility and high-rank state officials who were forced to have their residence in the city. Thus there appeared a class of people immediately interested in the environmental well-being of St. Petersburg and its outskirts where they lived in summer. Those high-rank people cared about the value of their possessions and did not want to inhabit an unhealthy place. Due to their capability of holding control over the state administration, the measures for canal draining, protection from floods, planting trees, etc., were possible even in the complicated political situation of eighteenth century Russia. They made planting pines incredibly large for that place, and some still exist in place of the former estates of rich people of St. Petersburg. Finally, the demand for food commanded laying out gardens in the surroundings which required melioration of suburban terrains. This highly enhanced the environment capacity of a large territory near St. Petersburg.

We can also cite an example of draining marshland in Adzharia on the eastern coast of the Black Sea, the ancient Colchis, which was almost uninhabitable in the end of the 19th century. Russian engineers accomplished a miracle of land-reclamation, a miniature of what the Dutch did for their country. The work was stimulated by Russian officials who received from the government estates in that land, sunny but unfit for living.

Finally, the carrying capacity of environment in a whole country such as Israel has been successfully increasing lately due to most advanced technology. Secular climate changes made a rocky desert out of a fertile land of ancient time. The country suffers most severely from the lack of water. Sharing the available water resources even became a matter of politics. The Israeli came to solve the problem otherwise than by the traditional use of force. Instead, they invented droplet agriculture in which water flows along plastic tubes to feed every single plant which helps avoiding evaporation from the whole irrigated area as it occurred everywhere in all times before.

The cited examples concern mostly land use and wise solutions of the water problem, either its excess or deficit. The solutions found for those historically older problems open also the prospects for increasing the carrying capacity of environment in terms of industrial wastes. Of first priority are studies of the dynamics of wastes removal in soil and maintaining the capacity of soil to remove different pollutants, for instance, heavy metals that reach the concentrations of hundreds MPC in large industrial cities. A known method to remove heavy metals, applied in some places, is to plant certain species of trees around the polluters. The plants absorb the heavy metals which thus pass to the foliage. It is then important to timely take away and bury or recycle the fallen leaves to prevent heavy metals from penetrating back to the soil. This is just one, wise and quite cheap, solution to the problem of pollution removal and maintaining the carrying capacity of environment. Economy and environment not necessarily should be irreconcilable antagonists. There are no reasons to close out plants and refuse the technological advance as the Greens claim. It is just important to bear in mind that economy cannot be isolated from environment and people should provide their balanced joint progress.

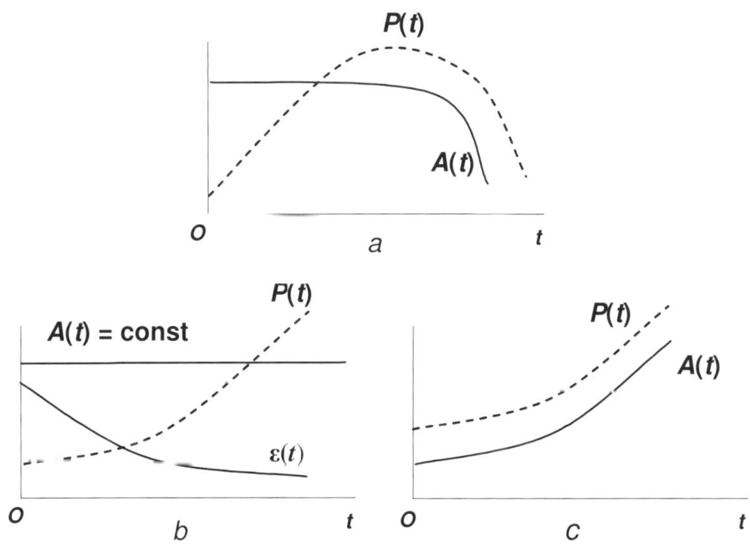

Fig. 10. Three possible evolution scenarios for economy and environment.

In conclusion we outline three possible evolution scenarios for economy and environment.

(1) Unlimited production growth with no care for environment leads to an environmental disaster and the ensuing cessation of production (Fig. 10*a*), which has already happened in some territories of developed countries.

(2) The balanced interaction between economy and environment (Fig. 10*b*) means cutting off pollution simultaneously with the advance

in production, without changing the carrying capacity of the territory. This scenario works in North America and most European countries.

(3) An ideal scenario (Fig. 10*c*) implies progress of production simultaneous with improving the carrying capacity of environment, even if pollution reduction fails, though this is desirable in any scenario. This is a harmonious coexistence of technology and nature, which can be called symbiosis. This end is quite possible if the environmental component of life is respected all around in all activities of man.

Note that a global symbiosis of economy and environment (especially, atmosphere) can be also achieved by reducing the overheating of the Earth by increasing its albedo. For this the possible technological ways can aim at maintaining the existing snow and ice cover in different territories as well as introducing light backscattering particles into the upper atmosphere, as was, for example, suggested in *Shaidurov, 2005*.

REFERENCES

Krylov M.P., 1995. Public response of population in Russia to the modern environmental situation, *Izvestiya RAN, Ser. Geogr.*, 6, 52–58.

Shaidurov V., 2005. *Atmospheric hypotheses of Earth's global warming*, University of Leicester, Mathematics and Computer Science, Technical Report No. MA–05–15, 8 pp.

Chapter 9

Market

In the previous chapters we discussed the normal and emergency behavior of complex natural systems, with a special focus on the impact of human activity on natural processes. Human activity occurs in complexly structured social systems of different levels. It is the dynamics of social structures that makes the subject of the following chapters of the book.

The existence of man and civilization primarily implies some mode of production and distribution of the products among individuals, smaller or larger groups of people, and finally among nations. This distribution is mediated by market in its different forms. Chapters 9 through 13 deal with problems related to the dynamics of market and possible catastrophes in economic systems.

Market and optimization of production

Market has a long history since the times of primitive pre-monetary barter trade, or even earlier when trade between tribes was very rare but each tribe had to solve its home "optimization" problem of achieving the best use of available resources. That was actually the problem of obtaining sufficient supply of any consumed product for the least labor cost by choosing the most efficient ways and refusing the others. Solving optimization problems of that kind was critical for tribe's survival.

To investigate the basic mechanism of production optimization, we first consider a tribe that lives on cropping and grows grain. At this point we assume that all grain is uniform in quality. The neglect of quality is fraught with serous errors but we apply this assumption for the starting simplest case. Then assume, again for simplicity, that grain grows in areas about the same size and each area is used by one member of tribe (and his family), and all grain producers are equal in efficiency and life standard. (We leave aside the privileges of chiefs and priests who were excused from productive labor for other purposes.) These conditions, which sound coming from the Utopian communism, did exist in most primitive grain-growing tribes that preserve till nowadays, such as the Pueblo Indians in Mexico and the southern US. Anthropologists who study those tribes believe that way of production was typical of all cropping cultures prior to the advent of private property. Note that the tribe members were only transient "owners" of their areas as the communal land was repeatedly redistributed, which was the usual way in peasant communities in many Indo-European nations.

Let the areas be labeled with the subscript i, n be the total number of areas suitable for grain growing, and P_i be the annual output of the i-th area (the P values can be different according to land conditions and crop capacity of areas). Then the total output of grain from all areas is

$$P = P_1 + P_2 + \ldots + P_n.$$

If the tribe counts N people and each person consumes P_0 pounds grain per year, the annual grain demand is

$$\underline{P} = N \cdot P_0.$$

This consideration is restricted to the case when $P > \underline{P}$, i.e., when all people can be provided with grain if all areas are in use. If the tribe does not care about the surplus product (as in the Pueblo Indians), only the most fertile areas are to be cultivated, and the optimization problem consists in selecting these areas.

Cropping requires labor input which differs in different areas according to their crop capacity. Labor cost is of course non-monetary as we discuss a pre-monetary society, but it is made up of the cost of tools and the labor itself. In the absence of money, these costs are expressed via the amount of product (its quality is assumed invariable). For instance, Japanese peasants paid all their taxes in rice till the end of the nineteenth century. The labor being assumed uniform, its cost can be tentatively measured against the amount of grain for which a tribe-mate would accept to do the job for somebody. Let it be S_i pounds grain from the i-th area.

We stress again that the *quality* of both grain and labor is neglected in the simple starting model, but below we look into this point missed in Ricardo's and Marx's theories of value.

We image the n areas, each with its efficiency (output) P_i and labor cost (input) S_i, using the phase portrait approach [*Khlebopros*, 1994][a]. Each area is shown in the phase plane S, P (Fig. 1) as the point (S_i,P_i). All points i = 1,2,..., n fill up a cloud contoured by a closed curve, and the task is to select the part of the cloud that gives the demanded amount of grain at the least total labor cost.

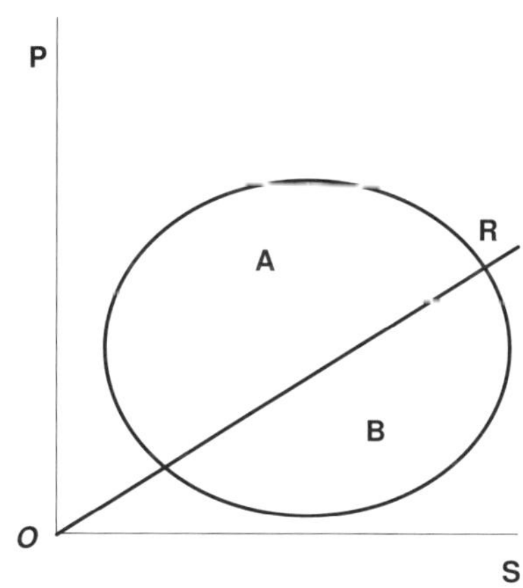

Fig.1. Distribution of areas economic (domain A) and non-economic (domain B) for cropping.

Let the set of areas that fit the choice be numbered as 1, 2, ... , m (m < n), then all grain demand is satisfied with

[a] The method of phase portraits apparently has never been applied before in studies of market and pricing.

$$P_1 + P_2 + \ldots + P_m = \underline{P},$$

at the least total labor cost

$$S_1 + S_2 + \ldots + S_m = S.$$

Labor input at the i-th area is S_i/P_i, which geometrically is the slope of Oi to the P-axis. Obviously, it is wise to use first the areas i for which the slope is the shallowest, and the areas are thus selected successively according to the slope. The straight line R that cuts off the selected areas is obtained by clockwise rotation of the line R from the axis OP (Fig. 1) to the position when the total output of the areas along and above R reaches the required value $\underline{P}$. The line R divides the cloud of the points i into the domains A (above R) and B (below R).

The areas that fall into the domain A are the most economic for grain growing, which is easy to prove. We assume that one area (i) in the domain A remains out of use and demonstrate that there is a more profitable way of using the available cropland. The total output $\underline{P}$ of all areas being fixed (this is the output of areas within the domain A contoured exactly according to this amount), the misuse of i should be compensated by the use of some areas from B. Suppose they are $j_1, j_2, \ldots, j_k$ and they yield the output P_i missed as a result of the misuse of i. As we show below, it is expedient to stop cultivating those and work only i instead.

Let j denote any area in the domain B. Then the line Oj slopes shallower to the S-axis than Oi, as the former is below R and the latter is above it (Fig. 1). The slope is measured by the P/S ratio which is the same for all points along the straight line, and for all points j

$$\frac{P_j}{S_j} < \frac{P_i}{S_i}$$

By multiplying both sides by S_iS_j and reducing, the inequality becomes

$$S_i P_j < S_j P_i.$$

These inequalities are valid for all j in our set, i.e., for $j = j_1, j_2, \ldots, j_k$. Summing them up gives

$$S_i P_{j_1} + S_i P_{j_2} + \ldots + S_i P_{j_k} < S_{j_1} P_i + S_{j_2} P_i + \ldots + S_{j_k} P_i,$$

or

$$S_i(P_{j_1} + P_{j_2} + \ldots + P_{j_k}) < (S_{j_1} + S_{j_2} + \ldots + S_{j_k})P_i.$$

Yet, the sum in parentheses in the left-hand side equals the output P_i of a single area i, according to the choice of the areas j which substitute i, and the sum in parentheses on the right is the total labor cost S in all areas j. Thus, $S_i P_i < S P_i$, or $S_i < S$, i.e., the substitution of a single area i for several areas j gives an equivalent output but a lower labor input and is thus more economic.

In a general case, several areas (rather than one) can remain misused in A and a number of areas from B can be used instead. In this case it is likewise easy to prove that it is profitable to refuse the areas from B and change them for areas in A. Note that the labor input is the least when only the areas from A are in use. Therefore, one can expect that the tribe after some trial eventually finds the best areas and abandons the others, having no need in extra grain.

Thus, the two-dimensional phase portrait method we applied reveals which land actually comes into use and explains the origin of the known wheat or corn belts, such as the Wheat Belt in the North American Great Plains where wheat is the dominant crop. Of course, no primitive tribe is able to apply exact mathematical calculation with regard to all input data (efficiency and labor cost). The task is difficult even for the today's society. The solution is rather of theoretical interest as it makes clear many economic phenomena. A tribe finds the optimum solution empirically trying and exchanging different areas, as it occurs in the above proof[b]. This is a remarkable feature of spontaneous economic processes, which repeatedly shows up in our story: being guided by personal incentive to prompt benefit, these processes eventually drive at

[b] The proof belongs to David Ricardo.

the optimum solution that fits the exact results of mathematical modeling.

The phase portrait of Fig. 1 is perfectly applicable to the modern conditions of private property and monetary economy.

Consider again growing grain but in terms of farming in a modern monetary culture where the areas i are used by permanent owners rather than by transient hands in a tribe. The output of each area i is again measured in weight units (say, pounds) and denoted as P_i ($i = 1, 2,..., n$) but the labor input S_i can be now measured in money units (say, dollars). The input S_i includes the cost of farmer's[c] working time S_{ib} (the price of a standard working hour is known for a given place and a given time) plus the cost of tools and materials S_{ia}. The sum S_i is equivalent to the money some imaginary "strange" owner would pay for working the area in the same way as the real owner does. The labor cost S_{ib} of the i-th owner is measured against the cost of work-hand hired for the fixed hourly pay:

$$S = S_{ia} + S_{ib}.$$

Assume that P_i, S_i and $\underline{P}$ (total annual grain demand) are known. Then we obtain a cloud of points i of Fig. 1 in the same way as above and draw the straight line R between the domains A and B so that all areas within A yield in total the demanded amount of grain $\underline{P}$ at the least labor cost.

This selection of crop lands may appear too "socialistic" as it aims at best satisfying public demands of the community, whereas the "capitalistic" choice under monetary economy and competition would imply the self-interest thinking of each producer. It turns out, however, that the private interests of individual producers drive at the same end as the optimum planning for the public good!

Suppose that the total output has reached the demand $\underline{P}$ (otherwise, the unsatisfied demand would call for using additional areas). If some area in A remains out of use, a group of producers who own the areas j_1, $j_2,..., j_k$ in B can trade-off their areas for i which provides the same output

---

[c] The common practice in farming is that a farmer works together with his family. For simplicity, by farmer's work we mean the work by himself and his family as a single labor unit.

at a lower total labor cost, as we showed above. In a monetary economy, this exchange means that they buy the area i from its owner having sold their uneconomic areas, say, to people who mean other uses of the land (e.g., other crops or pasture, etc.). The owners of $j_1, j_2, \ldots, j_k$ may loose in money but only once, whereas the earnings due to yearly gain in labor cost sooner or later compensate for the loss. The previous owner of i benefits as well because he does not use his land anyway and readily agrees to sell it for a good price. These transactions, in which both sides benefit, continue until the entire domain A becomes used for grain cropping and no grain production exists outside.

The previous discussion of grain market already assumed existing some market or at least the underpinning attitudes of market behavior. These attitudes are a subject of a detailed study below.

Prime cost

The prime cost of grain from the i-th area (C_i) is measured as labor cost (in dollars) per pound grain:

$$C_i = \frac{S_i}{P_i}.$$

C_i has a simple geometrical meaning (Fig. 2): it is the slope ratio corresponding to the angle between the line Oi and the P axis, or the S-to P-coordinate ratio of the point i (S_i/P_i). The highest prime costs lie along the line R because the largest angle POi for the areas i in A (shown as points above and along R) is POR (Fig. 2). If i_0 is an area along R, the prime cost of grain in it (C_0) is the S_0/P_0 ratio. This ratio is independent of the choice of the point on R: for any other area i_0' along this line, by similarity of the right triangles S_0Oi_0 and $S_0'Oi_0'$, we have

$$\frac{S_0'}{P_0'} = \frac{S_0}{P_0}$$

i.e., the prime cost of grain from i_0' is likewise C_0.

Thus, the highest prime cost of grain in the domain A is the slope ratio R/P valid for producers whose areas lie along the line R.

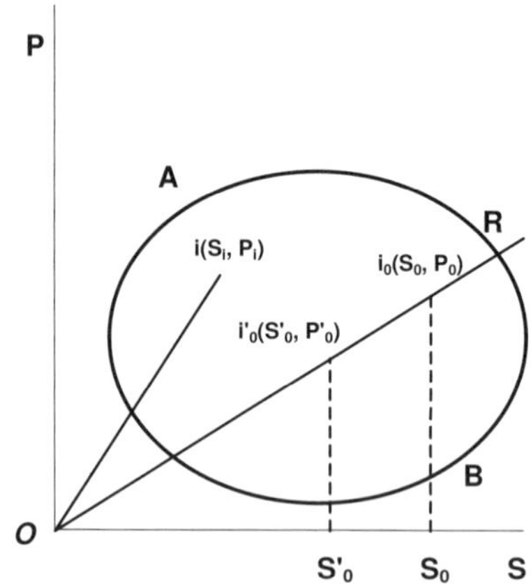

Fig. 2. Optimization of grain production in terms of prime cost.

Free market

The exchange transactions we referred to when illustrating the market mechanism in terms of optimization mean that the owners in both domains A and B know their efficiency (output) and labor input. Therefore, at the onset of market all areas should be in use; otherwise we assume that the cloud of Fig. 1 includes only the areas that have their owners and are in use when market begins. A modeler scientist who knows the values S_i, P_i and $\underline{P}$, can anticipate which producers fall into the domain A and will be successful in selling their grain and which will be unable to sell their grain and get ruined. However, the owners themselves remain unaware of this and just grow their grain, offer it at the market, and it is the market that selects which producers to keep and which to push out of business. As in other cases, the spontaneous market activity brings to the same end as mathematical prediction, which is actually impossible in practice for picking and processing all necessary data is too difficult.

Postulates of free market

What happens in the market?

The free market we discuss below is an ideal image of real markets. The numerous known concepts of free market imply primarily the voluntary informed transactions between producers (sellers) and consumers (buyers) without coercion and fraud and without intervention on the part of government or other organizations. We define free market through prohibitive postulates saying what is not allowed rather than what actually occurs, and put off for a while the problem of motivation in market agents.

1. No middlemen are allowed, i.e., producers sell only the goods they produce and buyers use themselves the commodities they buy.
2. No concert among sellers for rising prices is allowed.
3. No concert among buyers for lowering prices is allowed.
4. No interference into transactions is allowed, beyond the above three limitations.

The four postulates may appear strange at first sight, not because their sense would be vague but because of doubts whether they do work in processes commonly referred to as market trading. Therefore, we would ask the reader to look into the comments that follow before making his or her mind.

The four constraints almost never work in practice or work just partly. *Constraint 1* fulfills at local markets where farmers bring their produce and are themselves the sellers. In the US these markets are set up in the outskirts of cities. They are very popular for genuine products and the absence of parasitic middlemen, but they are open few days a week and are rarely accessible easily enough. Therefore, most commodities are sold in shops. Local markets always have been very common and were successful if remained free from the control of criminals.

Constraint 2 means ban against any monopoly. Concert among producers is usual in the modern world. It is especially pressing if there are few producers: buyers are forced to pay much being deprived of choice. Anti-monopoly laws were enacted in many countries already in the early 1900s and treated as unfair practice the monopoly of few (or even one) producer(s) who force out the others. Legal measures, however, are poorly efficient, leaving aside the moral side. It is only the competition that can really resist monopoly.

Constraint 3 bans boycott. Buyers can boycott a market if they are able to get the necessary commodities otherwise or can do without them for a long time. Yet, boycott is hardly practicable as buyers are too many and too weakly related to one another. Or, boycott can work if it follows some idea beyond economy, such as, for instance, the boycott of timber from the Amazon rainforest that withstands the injurious felling. Note that consumerism active against forgery and unfair publicity can be likewise classified as concert and breaks the laws of free market. Therefore, we assume below that sellers and buyers act freely and independently being guided uniquely by their own interest as they mean it.

Finally, *Constraint 4* implies, especially, that the government makes no attempt to intervene through taxes, subsidies, minimum wages, price ceilings, etc. In the case of command economy, where decisions regarding production, distribution, and pricing are a matter of governmental control, free market fails but becomes inevitably substituted by black or underground market[d]. Interference from outside can cause shortage of commodities, i.e., break the supply/demand balance. As in the case of grain market (see above), we assume that the supply is never less than the demand. Another important assumption is that each producer is free to deal with his produce at his will, according to the concept of private property: each grain producer, for example, is free in disposing of all crops from the area he owns. Besides, we assume that taxing, which is also a kind of outside interference, is proportional to the yield and taxes are collected prior to market transactions; thus P_i correspond to the after-tax remainder.

The concept of free market coined by Adam Smith in the second half of the eighteenth century made the cornerstone of the science of

[d] These phenomena far from the normal market concept are beyond our consideration.

economics. This is a necessary starting idealization, which is to be corrected and developed for particular cases, similar to the concepts of line or surface in mathematics, mass point and blackbody in physics, etc. Some existing definitions of free market include various ways of concert and fraud, such as monopoly (forbidden by our second postulate) or false information on quality of goods. Quality will be a subject of special consideration (see Section *Quality of Commodities* in this chapter) and another market postulate to extend the above four constraints.

Motives of competition

As we specified, the postulates of free market tell rather what sellers and buyers should not do than what they do actually. The motives of sellers and buyers require separate formulations. We describe them as "rules" which are a generalized observation of their average behavior.

The rules that control the behavior of market agents, who are assumed to take voluntary and independent decisions of charging prices (sellers) and purchasing (buyers), cannot work if the market is not free in the sense of the four postulates. Therefore, the latter are necessary conditions of the market behavior but they are of course insufficient for people to behave as they usually do. The actual behavior of market agents is driven by the incentive to gain. The classical scholars of political economics believed this motive to be intrinsic to the human nature.

Assume for simplicity that prices form during the first period of market operation which we call the "first season". Let the total demand for the commodity $\underline{P}$ (grain in our case) be satisfied in every season and be season-invariable. A season consists of shorter time intervals (say, days), and each seller is assumed to set the price for his product in the beginning of the day and keep it fixed all day long.

What are the motives that guide sellers in charging prices? In the beginning of the first market day prices are more or less arbitrary, but are never below the prime cost. Prices in the beginning of every following day are set proceeding from the prices of the previous day. Mind that the market agents are unaware of the exact values of the parameters P_i

(output) and S_i (labor cost) for all grain areas and $\underline{P}$ (total demand) which are assumed known for the modeler who draws a phase portrait with the domains A and B, like one in Fig. 1. Inasmuch as the domain A is the most economic for grain cropping and yields exactly the demanded amount $\underline{P}$, the scientist can, in principle, predict which producers will fall into A and have a granted market for their grain. In practice, of course, no economist can have exact information and any prediction is tentative. Nevertheless, to highlight the market processes, we assume this information to be available for the scientist and unavailable for the sellers. Each producer has only a rough idea of the output of his own area and the prime cost of grain, and is thus never sure whether he belongs to successful producers and whether his grain is marketable (or whether he belongs to the domain A in our model), especially because, according to our assumption, the supply exceeds the demand. This very uncertainty drives the competition.

The first rule is that no seller ever sells his produce for a price below the prime cost. Consider again a grain market. The labor input of the i-th producer consists of the costs of tools and materials and the cost of labor time. The latter, according to our assumption, corresponds to the hourly pay — fixed in a given locality at a given time — for the work of a hypothetical work-hand who would do exactly the same job as the farmer does. The work is assumed uniform and similar in all areas but different in duration, which exactly makes the difference between the areas. Finally, we assume that each producer knows (at least approximately) the output P_i from his area and the prime cost C_i of his grain. If he sold his grain for a price C_i' lower than the prime cost C_i, his revenue $C_i'P_i$ would be lower than his labor costs $S_i = C_i P_i$:

$$C_i'P_i < S_{ia} + S_{ib}.$$

If the expenses S_{ia} are supposed inevitable for maintaining the cropping business, the remainder for the living expenses of the farmer (with his family) would be below S_{ib}, i.e., his life standard would be lower than if he had been a work-hand, which, of course, contradicts his expectations, i.e., ambitions and habits[e]. A producer can, for a while, work at a loss by a prestige of being an owner, but we leave aside this exception and assume that prices are never below the prime cost.

Another rule represents the competition attitude of sellers. Knowing the prices of the previous day a seller bewares of charging a price above or even equal to the lowest previous price not to loose the competition with others who would keep the price of the previous day. Therefore, he charges a price slightly below the lowest previous price provided it is above the prime cost of his product. Otherwise the price of his grain equals its prime cost which allows the seller to have the least number of competitors and wait till the more successful sellers who can afford lower prices sell out their goods and leave the market.

The above motivation of sellers is guided by the behavior of buyers who always buy for the cheapest possible price. Mind again that our simplified model neglects quality and assumes the price to be the only motive for buyers. Another rule of the buyer's behavior is that if prices are going down, buyers wait until the fall stops and thus save by buying the cheapest commodities. Of course, these rules not always work in practice. Moreover, it is in the first market days that price cutting is induced by the fact that some commodities are already being bought, for the cheapest prices. Yet, our simplified assumption is that most agents behave according to these rules. To sum up, they are as follows:

1. Seller never charges prices below the prime cost of his product.
2. Since the second market day, seller charges prices below the cheapest price of the previous day if it is above the prime cost of his product or equal to this prime cost otherwise.
3. Buyer buys for the cheapest price that exists in the market at the moment of purchase.
4. In the case of price cutting buyer does not buy before the decline stops.

---

[e] This is a social element in the motivation (see the details below), whereas S_i is defined by objective economic conditions.

Pricing mechanism

Now we can proceed to the mechanism of pricing. In the first market season, the supply always exceeds the demand (according to our assumption of no shortage). Sellers are unaware of the total demand $\underline{P}$ and of the conditions of their competitors and are uncertain about their market success, which drives the competition. The model of pricing we suggest is of course very simplified and highlights only the basic behavioral tendencies of standard market agents.

On the first market day, prices can be more or less arbitrary though never below the prime cost for each seller (Rule 1). Since the second day, each seller cuts his prices down to some level below the cheapest price of the previous day if the latter price is above the prime cost of his product or to this prime cost otherwise (Rule 2). The price cutting is induced by purchasing for the cheapest available price (Rule 3), but these purchases are few as most buyers wait until the decline stops (Rule 4). We neglect the few exceptions and assume no purchasing during price decline.

Finally, the price fall stops when all sellers fix the prices equal to the prime cost of their products. (Note again that the price charged by the i-th seller is shown in the plane (S, P) as the slope of the line Oi to the P-axis, see Fig. 2). Then the buyers begin buying until the seasonal demand $\underline{P}$ becomes fully satisfied. They buy for the cheapest available price (Rule 3), i.e., select the seller whose line Oi rises the highest. Thus all goods are sold for the prime cost in the first season.

The modeler means that the producers from A have fully satisfied the grain demand and can check the prediction of who will have the market according to the line R in the phase portrait. However, the producer who sold out his grain likewise knows he will have the market in the following season because he saw that people bought from others after he had sold out his grain and that they bought for the price above his prime cost, which is the evidence of unsatisfied demand. The producers who were not among the last sellers can expect that, should the demand hold, their grain will be wanted in the following season as well and they won't go broke. (They make sure to belong to "elite" in the terminology of the

Genevans, the first partisans of free market, or to "the domain *A*" in terms of our model.)

Therefore, in the second season and in all seasons on, sellers charge the highest price for which grain still found market in the previous season, i.e., the highest prime cost C_0 in the domain *A* which is defined geometrically by the *R*/*P* slope ratio. Why don't the producers, who sold their grain in the first season and are sure it is demanded and marketable, charge prices above C_0? The reason is that there remain some sellers whose grain has the prime cost slightly above C_0 and who have not left the market in hope to get right in the following season. Therefore, competition holds in the second season: Should the sellers successful in the first season charge the price C^* above C_0, they would have to compete with the sellers who own areas along the line *R* in the domain *B* and can charge the price above their prime cost but below C^*. Thus it is the existence of the least successful producers who strive for survival that hinders further price growth. C_0 is thus the fixed price of a market, the grain market in our case. Obviously, it changes in response to changes in the distribution of areas or in total grain demand.

Turning back to the postulates of free market and their relation to the market behavior, note again that the postulates are the necessary but not sufficient conditions that uniquely define this behavior. There is an approximate analogy with physics. The conservation principles in physics likewise look as prohibitions: 'the energy of a system never disappears or appears from nothing', 'momentum never changes', etc. Yet these principles do not determine the dynamics of the system described by different laws related to the causes of motion. All conservation principles work in a moving system but cannot describe its motion. Of course, they are valid only for closed systems not subject to outside impact. In economics, free market is similar to such a closed system.

The economic analysis of actual market-related facts is free from any moral judgment about the rules of the market behavior. The aspiration of selling at highest and buying at cheapest is unworthy of the lofty ambitions of a gentleman. Yet we investigate below how a gentleman could get his rent which encouraged that elitism. Furthermore, the market behavior is at odds with the Christian attitudes as they are read in the Sermon on the Mount and the attitudes of socialists and the Russian intelligentsia derived therefrom. Thus, many readers may dislike the market thinking. Nevertheless, the effect of free market with the motivation behind the behavior of its agents, though associated with disagreeable thinking, is to improve the living standard, at least for the community as a whole if not for all its members.

Rent

We again invoke grain production to approach the concept of rent. Let the producer i have the input S_i and the output P_i and sell his grain for the fixed market price $C_0 = S_0/P_0$. Since $P_i = P_0$ (Fig. 3), $S_0 = P_0C_0 = P_iC_0$, and C_i, the prime cost of one pound grain from the area i, is $C_i = S_i/P_i$, wherefrom $S_i = C_iP_i$. The length of the horizontal segment between i and i_0 is $S_0 - S_i$, given by

$$S_0 - S_i = P_0C_0 - P_iC_i = P_i(C_0 - C_i).$$

The segment ii_0 measures the labor input difference between the i-th area and the "worst" area i_0 which has the same output but a higher prime cost of grain. According to the market price equation, $S_0 = P_0C_0 = P_iC_i$. The value S_0 is nothing but the total revenue of the i-th producer from selling P_i pounds grain for the market price C_0 dollars a pound, and S_i is his total input into growing his grain. Thus $S_0 - S_i$ is earnings due uniquely to the fact of owning good land. Indeed, the producer that owns the inferior land of i_0 has the same grain yield ($P_0 = P_i$) but at a greater labor cost. The difference $S_0 - S_i$ is the *rent* of the i-th area, written as DS_i (D stands for 'rent' and is not a product factor), which corresponds to the segment ii_0 (Fig. 3).

Now imagine that the owner of the i-th area does not use it any longer but leases it to the owner of the area i' from the domain B (Fig. 3) who cannot live on his crops as he gets a lower output $P' < P_i$ at the same costs $S' = S_i$. Indeed, the owner of i' has to sell his crops for the fixed market price C_0 and his gain $P'C_0$ is lower than the gain from the area i and even lower that the gain of i_0', the least successful producer from A running the same expenses for whom $C_0 > P'C_0$ (Fig. 3). As shown above, producers from the areas along the line R sell their grain for prime cost which makes the market price C_0, specifically, the prime cost for i_0'

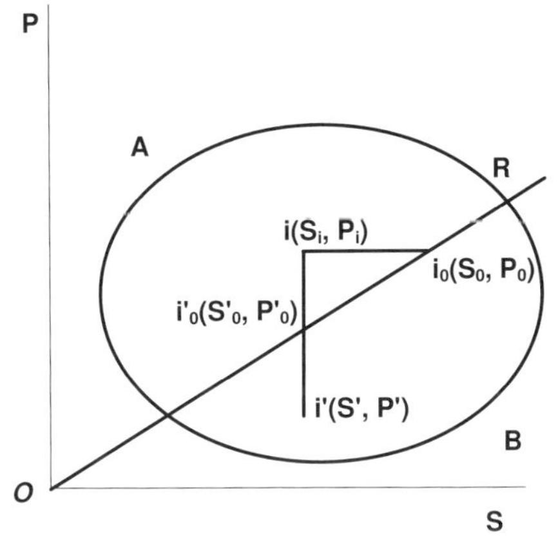

Fig. 3. Formation of rent.

is $S_0'/P_0' = C_0$. Compared to these, grain from i', has the prime cost S'/P', and $S' = S_0'$ but $P' < P_0'$, and

$$\frac{S'}{P'} > \frac{S_0'}{P_0'}$$

(the right-hand side has the same numerator but a greater denominator). The right-hand side of the inequality being C_0, grain from i' has its prime cost above the market price, i.e., even the labor cost is not recovered! Therefore, the i'-th producer will readily agree to rent the area i for the annual revenue $S_i = S'$ which is equal to his input into i'.

According to the contract, he will have the same revenue as the owner of i_0' who works for the prime cost of his grain like all owners along the line R because $S' = S_0'$ (Fig. 3); but as $S_i = S_0'$, the tenant of i will put in the same labor cost as the owner of i_0', i.e., the former owner of i' finds himself, as a tenant, in the same conditions as the least successful owner i_0' at the efficiency limit of the domain A.

Yet, the earning from the area i is actually $P_iC_0 = P_0C_0$ (since $P_i = P_0$), which is S_0 because it follows from $C_0 = S_0/P_0$ that $P_0C_0 = S_0$. This sum includes the fixed revenue of the tenant S_i and the rent $DS_i = S_0 - S_i$ he has to pay to the owner of i.

Therefore, whether the tenant agrees or not to pay the rent DS_i depends on the above inequality meaning that the prime cost exceeds the market price. The prime cost is S/P and the previous inequality means that

$$\frac{P'}{S'} < \frac{P_0'}{S_0'},$$

i.e., the slope of Oi' to the S-axis is lower than the slope of R. Thus the owner of i can offer it for lease only to producers from the domain B. Inasmuch as this inequality approaches equation in the vicinity of the divide line R, the owner of i can claim a higher rent only from very unlucky producers who are far from R. Therefore, the rent $DS_i = S_0 - S_i$ is the granted least gain from the lease of the i-th area.

Thus, the owner of i who has leased his area (for some period according to the contract) can have his rent for nothing doing. Of course, a typical landlord owns many areas covering a large territory and can live decently on the rent. This is how the exploitation of man by man originated with the origin of rent. Exploitation arises as soon as somebody works using land or factory, or any other means of production not being their proprietor while the legal owner put in no labor but gets a percentage of the gain from the sales. The parasitic behavior of landowners who make no effort to improve their land viewed against the fully responsible tenants looks especially outrageous. Some landlords may play an important part in management if they control the state of the leased land, themselves or via a hired manager, and refuse the lease to bad tenants. In many countries (e.g., in Latin America and Russia), landlords often leased their land to share-croppers and were only interested in the highest current income. Then both, a bad tenant and a careless owner could be equally short-sighted in abusing land.

This is the way feudal landlords got their rent long before the onset of capitalism, and the rent size depended on the crop market. This way is feudalistic though it stems from market relations. A tenant exploited by a landlord lives in better conditions than a wage laborer as he is a decision maker, he is free in using his time, and he lives with his family who help him in work, though most often the house where he lives belongs to the landlord. This independence ensures a sounder life that better fits human dignity than the life of a wage worker at a capitalistic factory (see Chapter 11).

We put aside the arguments blaming the parasitic way of landlords but excusing capitalists as "managers". Some land owners do act as capitalists who manage the production. In our simplest case of grain growing, the "capitalist" does not lease his land, makes decisions himself but uses work-hands. The work-hand labor is cheaper than the labor of a responsible tenant. The work-hand pay is of course included into the labor cost; the owner gains his rent, which is lower than if he worked himself but higher than if he leased his land, otherwise he would have no motive to hire work-hands instead of leasing.

Now imagine that somebody who wants to earn from cropping buys the area i to lease it. Let p be the mean profit rate in farming, i.e., the mean annual return from investment into farming is $p\%$ of the input capital. Then the buyer of the area will expect that his investment, or the price C of the area, provides exactly the same return, or $(p/100)C$ dollars per year. This amount is the rent of i, or

$$DS_i = (p/100)\cdot C,$$

whence

$$C = DS_i\cdot 100/p.$$

This is the mechanism of land pricing. The today's rate of profit is 7–8% in farming and as high as 10–11% in industry. The question then arises why people invest into croplands. The reason is that the return from factories is less safe and of a shorter run because the costly upgrading is required every eight to ten years. Another motive is that people want to leave a reliable source of revenue to their children. This incentive, which used to be very strong in old times, is now restricted to backward feudalistic economies and ever more often gives way to the egoism of the current generation (see Chapter 14).

Population change

Hitherto, we have assumed that the consuming population remains invariable. If population grows, the increasing demand requires using a greater number of farming areas (the cloud of points in Fig. 1) and, hence, rotating the line R through a greater angle. This increases the ratio $C_0 = S_0/P_0$, i.e., the market price for grain; population decrease, on the contrary, causes price fall. The effect of population growth can be illustrated by classical examples, beginning with Robinson Crusoe. He had several areas good for cropping in his island and, of course, selected the best ones to put in the least labor per unit product. If he made the phase portrait as in Fig. 1, he would draw the line R to have good areas with the necessary output above R. When Friday arrives, they have to work more and the line approaches the S-axis. However, labor productivity increases with population growth due to cooperation, i.e., labor cost per unit product S/P decreases (Fig. 4)[f]. If population grows too fast, people have to use ever worse land, the prime cost of products again increases, and the life becomes harder.

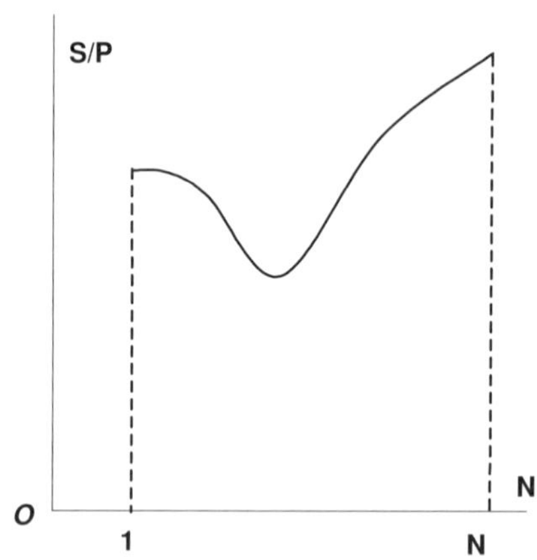

Fig. 4. Dynamics of prime cost as a function of population dynamics.

[f] Note that Fig. 4 refers to the situation when population growth is not accompanied by breakthrough in technology as a result of inventions and discoveries.

In animal communities, rent corresponds to energy gain. A growing herd of horses or buffaloes grazes up the herbage sooner and has to move to inferior lands, which takes them more energy to feed. On the other hand, a greater population is advantageous in defense against predators. Birds and insects, defenseless one by one, join into flocks to defend themselves together. Konrad Lorenz explained it suggesting plausible mechanisms.

Locust invasion is a spectacular example of expanding the feeding territory of a population. Locust has been found out recently to be a species allied to grasshoppers (see Chapter 1). If the density of a locust population in some habitat – say somewhere in Mongolian steppes – surpasses a certain limit, the available territory becomes insufficient to feed the population. Then the insects develop wings, which are normally reduced, and gather in flocks able to travel over enormous territories devouring all vegetation; then they fly off to turn back into common "grasshoppers". These biological phenomena are driven by economic causes, namely striving for the best use of available territory, that is for increasing the rent.

Thus the difference in input per unit product has been important already in animal communities and had existed in human communities long before the appearance of landlords; it makes the basis of rent. The maximum rate of profit can increase as new types of production come into use to maintain a greater population in the same territory. This is one possible explanation of why the groups of earliest humans gave way to tribes and then to million-scale nations, such as Egyptians or Chinese.

Landlords getting their rent stir up resentment in their recent "tribe mates" who treat them as parasites. Landowning is, however, advantageous as far as landlords care about improving the land through the appropriate land use and refusing bad tenants. An alternative way of production that fits the tribal ideal of social justice consists in communal land holding without private ownership but with repeated redistribution of land, whereby rent becomes spread over the community and everybody gets his fair share. This way was common to all Indo-European peoples and long coexisted with landowning: somebody could obtain "legal" ownership of land by conquest or grant from his sovereign and the existing tribal community became a "cooperative user" which paid rent to the owner. Land-use communities are known from medieval Germany (so-called *marks*), have persisted in India till nowadays, and existed in Russia as long as the 1900s[g]. The disadvantage of communal land holding is that rent sharing makes peasants careless about the quality of land as its redistribution does not allow them to benefit from improvements. Therefore, communal land becomes ever worse whereas the land used by landlords via sharecropping or work-hand labor can ameliorate. Today's farmer holding is the most advantageous way, where each owner uses his land working with his family or few work-hands. This system roughly corresponds to Figure 1 under the assumption of the absence of parasitic landlords.

Quality of commodities

Now it is time to discuss quality of commodities which remained excluded from our previous modeling. One and the same kind of products that satisfy the same demand can highly vary in quality. There are eight hundred quality grades of grain! The neglect of quality is a serious drawback in early economic theories, for instance in that of Marx. Marx assigned some average quality to every product, including labor, which he treated as a commodity following Ricardo. According to Marx, the cost of commodities is measured by "average labor time" required for their manufacturing — the approach that cut off difference in skills and care. Perhaps, it was because Marx might have been stuck with the quality problem that he never finished the third book of his "*Capital*" which Engels published from rough copies after his death. The faulty approach in which all estimates were averaged without regard for their real scatter may have been the reason why the applications of Marx's theory failed, especially his concept of class-based thinking and behavior.

The model without quality shows that both non-market and free market ways of production optimization can achieve the goal of meeting

---

[g] The Russian peasant communities ("obshchina" or "mir") were once believed to be specific to Russia, which underlay the sociological theories of a "special Russian people".

the demand at the least cost: it is the optimization providing the least prime cost per unit product. This is the key point in Adam Smith's theory. However, once quality becomes included, free market optimizes a different parameter, as we show below.

What is quality? Historically the term was coined by A.N. Krylov in the latest nineteenth century for technological applications and was developed by John von Neumann who related it to the concept of utility.

The utility of a commodity can be estimated using non-monetary barter exchange and includes its properties as well as the demand and preferences of possible consumers. To estimate the utility ratio of two commodities that satisfy roughly the same demand, one can offer consumers to exchange one for another. If, for instance, there are whisky and beer in the market, buyers may accept to have three bottles of beer instead of one bottle of whisky, then a bottle of whisky has a utility of three bottles of beer. Of course, an individual buyer may have his special likings but consumers taken in total have consistent preferences which validates the correlation. Commodities of different kinds are more difficult to correlate, but von Neumann suggested to apply numerical trade-off equivalents (we mean non-monetary barter trade, without the concept of price). Thus, relative utilities of commodities can be measured on some scale against the utility of a "standard" commodity traded-off for various other commodities. For instance, with a bottle of whisky as the standard, the utility of a beer bottle is 1/3 and the utility of a TV set is, say, 1000. There, however, arises a problem of "transitivity", as the numerical utility values are different for a different standard commodity. Von Neumann presumed that the utilities corresponding to different standard commodities are usually consistent within a certain group of consumers (an ethnic group or a country). Therefore, a utility unit can be equated to a certain amount of any commodity, in the same way as units are chosen in any measurements.

What is commonly meant by "quality" is closely related to utility but is not its exact equivalent. For instance, whisky of any high quality would be of a very low utility in a Moslem country where nobody buys alcohol drinks (even for resale, as the sale of alcohol to non-Moslem people is punishable severely). This market is not quite free because of the legal limitations and, being confined within a country, soon pushes the unpopular commodity out of sale. The basic assumption in our further consideration is that the quality of commodities offered for sale in free market is equivalent to their utility. In a general case they are not equivalent, and quality is a subject of a sophisticated research (quality control), but this distinction is of no importance in our simplified model of free market. So we interpret quality of a commodity as its utility estimated according to von Neumann through non-monetary exchange. The quality and the market price of a commodity are measured independently and the relation between the two is never known *a priori* though appears obvious. Constraining this relation is a key objective of free market. Correspondingly, we include the following assumption into the definition of free market: the price of a commodity is proportional to its quality.

Inasmuch as we now study pricing with regard to quality, our definition of free market becomes extended with the fifth postulate[h]:

5. No seller is allowed to overestimate the quality of his goods or to charge higher prices for inferior quality.

The numerical estimate of quality equated to utility and measured through von Neumann's exchange refers to a unit product (say, a pound) rather than to product as a whole. There the scientific usage of the term *quality* diverges with the everyday meaning. If, say, a bottle of whisky is taken for a standard commodity, and beer is measured in bottles, the quality of beer as a commodity is 1/3. Thus, saying *the quality of a commodity is Q* means that a unit of this commodity (e.g., a pound grain) is traded for Q units of the standard commodity, or each pound of grain contains Q quality units.

Now we again model the grain market, but with regard to quality. There are again n areas i of grain cropping, with their outputs P_i (in weight units) and labor costs S_i, and the quality of one pound grain Q_i[i].

---

[h] Again, we study an ideal market which approximates the actual market (biased by government control, publicity, sellers' or buyers' concert, etc.) in the same way as line or surface approximates the geometry of an actual body or an ideal particle approximates a moving object of any size.

[i] Mind again that according to von Neumann's utility measured by trading-off, quality is

Note that grain from different areas is no longer the same! The prime cost of one pound grain from the i-th area is

$$C_i = \frac{S_i}{P_i}.$$

Now each point i in the plane (C, Q) with the coordinates (C_i, Q_i) has three parameters (Fig. 5): Q_i, S_i, P_i, and the coordinate $C_i = S_i/P_i$:

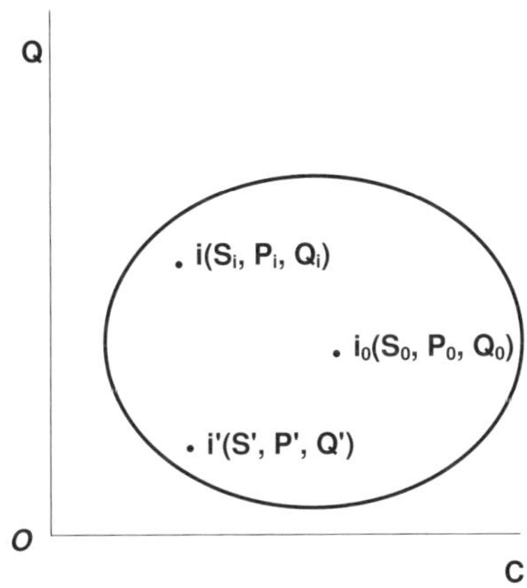

Fig. 5. Market of cropping areas that differ in grain quality, crop capacity, and labor cost.

As in Fig. 1, the points make up a cloud showing all producers in the market, each having the respective output P_i. The total output of all used cropping areas is

$$P = P_1 + P_2 + \ldots + P_n.$$

If the per capita consumption of grain is P_0 (in weight units), the total demand is

$$\underline{P} = N \cdot P_0,$$

where N is population. As in the previous modeling, we assume that the demand is satisfied, i.e.,

$$P > \underline{P}.$$

---

proportional to the quantity of the consistent commodity, i.e., the utility of two bottles of whisky is twice the utility of one bottle, where quality is equivalent to utility.

The areas economic for grain growing are selected proceeding from the criterion of the least prime cost of unit quality. Before we do the choice, consider two cases of non-market distribution of products. The first one was historically practiced in different world economies, including Soviet Russia.

Imagine that we are to feed the population in the cheapest way, which was the objective of the government of England during the Second World War and the Soviet planning offices in the time of "five-year plans". Then, neglecting quality, we move the vertical line V in Fig. 6 rightward from the Q-axis until the number of the points i on the left of the line and along it becomes such that the total output fits the demand $\underline{P}$.

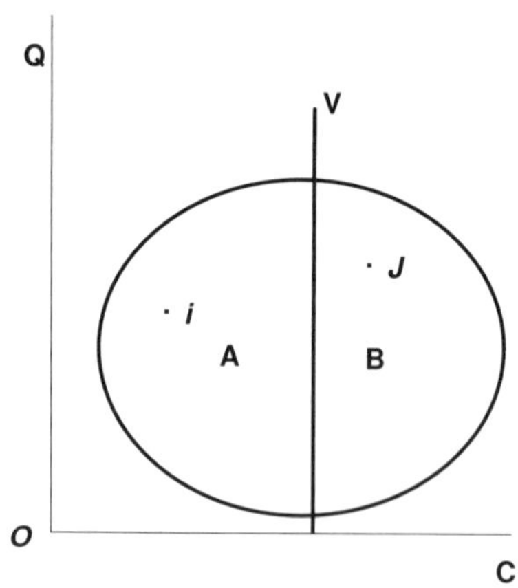

Fig. 6. Optimization in grain production: achieving least labor cost per unit product.

These points make up the domain A including the areas that yield cheap grain and the areas on the right (domain B) with more costly grain. The objective of feeding the population in the cheapest way, i.e., at the least labor costs, requires using only the domain A and leaving the areas of B out of use. This was the planning strategy during the war time in England and in the five-year planning in Soviet Russia, of course, not only for grain. The price of grain C_0 (if it is sold for the prime cost, or for the total labor costs for the government) is obviously

$$\underline{P}C_0 = P_1C_1 + P_2C_2 + \ldots + P_mC_m,$$

where the right-hand side is the total labor input in all areas from A and the left-hand side is the total of money paid by all consumers for the grain they buy ($\underline{P}$). The sum of all P_i over the points in A is exactly $\underline{P}$, according to the choice criterion for the areas of this domain. It is easy to derive from the equation that C_0 does not exceed the highest prime cost C_i in the right-hand side. Of course, the equation we used to find C_0 presumes that the price of grain is equal to and not above its prime cost, otherwise we would have an inequality with the sign >, and the state officials who fixed the price would spend the difference at their will.

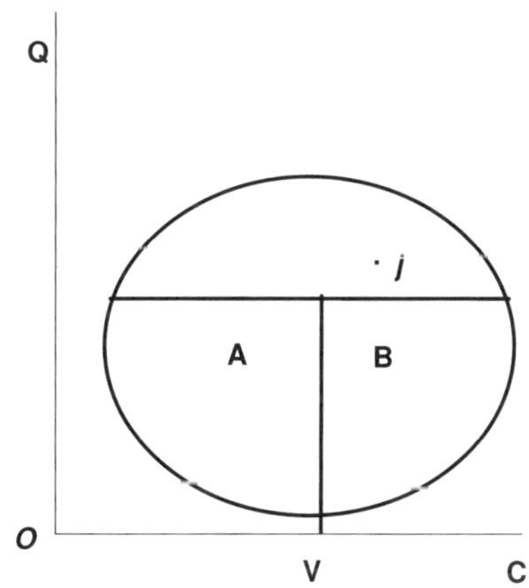

Fig. 7. Free-will of elite in non-market distribution of product.

The most curious thing in "socialistic" economies of this kind is that the uniform price for cheap product (C_0) is by no means related to its quality Q. However, the consumers do care about quality and would run all tricks and turns to "reach" the highest-quality product for this price. Grain from these "elite" areas j (Fig. 7) certainly goes to the officials responsible for the distribution of products, as an additional free grant for their official position. Although the objective was to feed everybody at the least labor cost, a "socialist" economy is not very effective in saving labor. Higher-quality products better satisfy the demand: one can wear a pair of good shoes for years and bad shoes wear out in a couple of weeks. Production which actually neglects quality aiming only at cutting current labor costs gives high output of bad products. As a result, the demand remains unsatisfied though much work is put in and much product is put out. If quality of a commodity is proportional to its ability to satisfy some demand, it is free market that fits the best the idea of meeting public demand at the least labor cost. Thus both free market and the appropriate optimum planning, which hardly can be put into practice (see Chapter 13), drive at the same ends.

Returning to the model of a "socialist" economy, note that in practice the officials make a more artful choice than we originally assumed. First, they just take the best product for themselves, as much as they wish for the price they fix, at no care about the labor costs. Geometrically, the i cloud becomes truncated as in Fig. 7, the top corresponding to the product caught by officials.

Then they release the remainder into sale proceeding from the principle "to feed everybody in the cheapest way", and leave their less privileged compatriots to their own choice of arts in getting higher-quality products. Therefore, all points in the top are eliminated and the procedure of Fig. 6 is applied to the truncated cloud.

Thus, public control over the activity of officials is especially crucial in the case of non-market distribution of some services (such as free education or public health care, etc.) because there arises a danger that minimization of costs per unit service calls forth a decrease in quality. This is now the case in many large economies.

Now consider free market, again its ideal approximation (without pressing from publicity or other factors, without shortage problems, etc.), which follows the four postulates plus the postulate that price should be proportional to quality. The way how free market, with the five constraints including the price/quality correlation, solves the optimization problem of achieving the cheapest unit quality is illustrated with a simple model below. The modeling strategy is the same as in the case of invariable quality.

We start again with Fig. 5 showing the cloud of points i economic for grain cropping, with the output P_i (in weight units), the quality Q_i, and the labor cost S_i (in money units, say, in dollars). Then the prime cost of one pound grain from the i-th area is

$$C_i = \frac{S_i}{P_i}.$$

The areas economic for grain growing are selected in the plane (C, Q) by clockwise rotation of the straight line V from the axis Q towards the axis C until the total output from all areas i above and along V reaches the total demand $\underline{P}$ (again, we assume that $P = P_1 + P_2 + \ldots + P_n > \underline{P}$). The respective points i make up the domain A and the remaining points j belong to the domain B.

The pricing mechanism is as follows[j]. In the beginning of the market, people buy for the prime cost C_i dollars a pound, and everybody looks for the seller with the greatest Q_i/C_i, i.e., seeks buying the best quality per dollar.

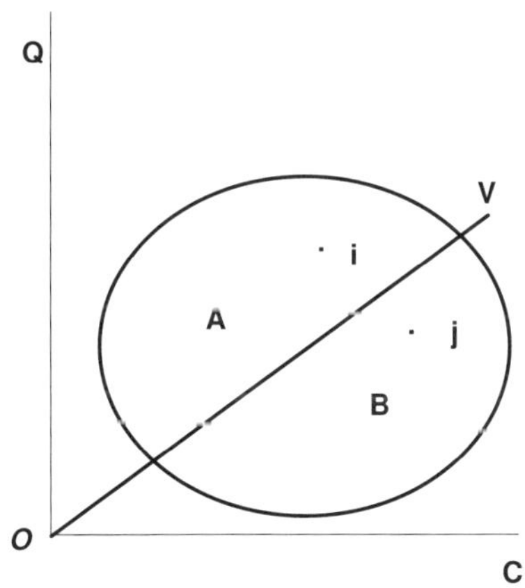

Fig. 8. Optimization in grain production by an ideal market: achieving least labor cost per unit quality.

However, the ratio Q_i/C_i being the slope of the line Oi to the axis C, buyers first seek the points i with the highest possible Q/C in the cloud. The buyers that come later to the market have to be content with grain from the areas where Q/C is lower and, finally, when all grain has been sold out, the latest buyer has it for the price corresponding to that for areas along the line V, i.e., equal to the prime cost C_0 at an area of this

[j] It is similar to that for the above case of invariable quality, and details are thus omitted.

line (let it be i_0). Then the price of unit quality for this buyer is C_0/Q_0, where Q_0 is the quality of grain from the area i_0, which is set from this point on as the price of a given market (Z_0):

$$Z_0 = \frac{C_0}{Q_0}.$$

Thus, the difference in optimization between non-market and market economies is that the latter seeks for the least prime cost per unit quality and the former can, in principle, choose any economic parameter for optimization. This is the reason why non-market ways of production and distribution survive and can coexist with market economies, in spite of their obvious drawbacks and hidden risks.

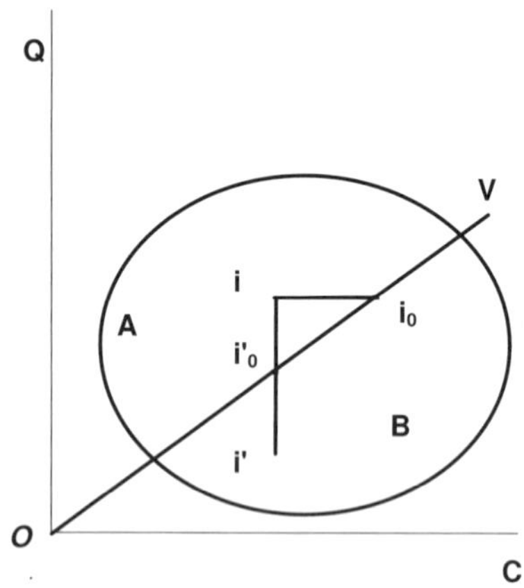

Fig. 9. Origin of rent and quality.

Now it is pertinent to revert to rent and investigate what can be its other controls but the prime cost C. We assume that there are two areas in the plane (C, Q), i in the domain A and i' in the domain B, with the same grain output but a lower quality at i' (Fig. 9). The owner of i' would most likely accept to rent the area i for the profit due to grain quality higher than in his former area i'. Indeed, according to the contract, the tenant of i will sell grain from i and leave for himself the equivalent of the grain cost from the area i_0' which lies at the intersection of the vertical $i'i$ and the line V (Fig. 9), where the quality of grain is higher than in i' but lower than in i. Then the tenant gets into the same conditions as the least successful producers from A, such as the owner of i_0'.

For all areas along the line V, the ratio C/V is the same and equals the market price of the unit quality Z_0. Specifically, at the prime cost C_0' and the quality Q_0' of grain from i_0',

$$\frac{C_0'}{Q_0'} = Z_0.$$

wherefrom $C_0' = Z_0Q_0'$. On the other hand, $C_0' = C' = C_i$ (Fig. 9) and $C_i = Z_0Q_0'$, which is nothing but the gain of the producer i_0' per pound of his grain; the tenant of i gets the same according to his contract. Yet, the total gain per pound grain from i at the quality Q_i is Z_0Q_i. If we draw a horizontal line through the point i until it crosses V at the point i_0, the quality Q_i at the point i is Q_0, and the total gain per pound grain from i (Z_0Q_0) is just the prime cost C_0 at i_0 (see the definition of price Z_0). Thus, the owner of i gains C_0 dollars per pound grain and gives up C_i to the tenant. The difference $DC_i = C_0 - C_i$ is the rent from the area i or the money the owner takes for his own from each pound of grain after the tenant takes his contracted share C_i. Geometrically rent is shown as the segment ii_0 in Fig. 9.

Figure 9 demonstrates the possibilities to increase rent: either to reduce the prime cost of products (move the point i leftward) or to increase quality (move the point i upward).

The Japanese chose the latter way after 1945. They decided to improve the quality of commodities using their cheap high-quality labor and the many skilled engineers who, during reconversion, quitted the war industry where the quality standard was higher than in the civil production.

The Japanese used a great part of the remaining currency holdings for buying licenses and bought the most advanced available information (instead of repeated simulation of the foreign know-how) and continued the progress from the point it has reached in the advanced countries. License sales is likewise a kind of rent, and the Japanese paid that rent to their victors. As a result, they achieved high quality, along with low prime cost, in many industries, such as ship-building, electronics, car production, etc. They used the earnings for reconversion, for rising the living standard of the population (and achieved five- to six-fold increase in fifty years) and for environment remediation. Remediation measures were pushed up by the fact that land was becoming ever cheaper because of pollution, for there was no place in the small country where the ruling elite could escape. Then, large-scale forest plantation in highland terrain started as early as since the 1955-60s.

To sum up, there are three ways of the origin of rent:

1. Reducing costs;
2. Improving quality;
3. Increasing crop capacity or production output.

Rent can be evenly shared among community members (as in communal land holding by peasant communities) or come into possession of individuals. It can go to the owners of land or factories, or also to the owners of intellectual property (this is the way the market of intellectual products such as projects, copyright, etc. can arise).

Social revolutions are thus nothing but the attempts to change the owner of rent.

REFERENCE

Khlebopros, R.G., 1994. Socio-economic estimation of ecological objects (Two limiting cases of the hierarchy of characteristic times), in *Global and Regional Ecological Problems*, Krasnoyarsk.

Chapter 10

Marketing Dynamics

The usual behavior of selling and buying agents in free market eventually drives at setting the highest price of goods. The phase portrait of this behavior shows sellers as the points i making up a cloud that fills some domain in the plane (C, Q), where C is the prime cost of a quality unit of goods and Q is quality (see Figs. 8 and 9 in Chapter 9). The coordinates (C_i, Q_i) of the point i in Fig. 1 correspond to the prime cost and quality of commodities offered by the i-th seller. For the points i_0 on the line V, which has the same sense as in Chapter 9, the ratio $Z_0 = C_0/Q_0$ (or the slope of V to Q) is the fixed price of goods, i.e., a quality unit in a given market is eventually sold for a price close to Z_0 but never exceeding Z_0.

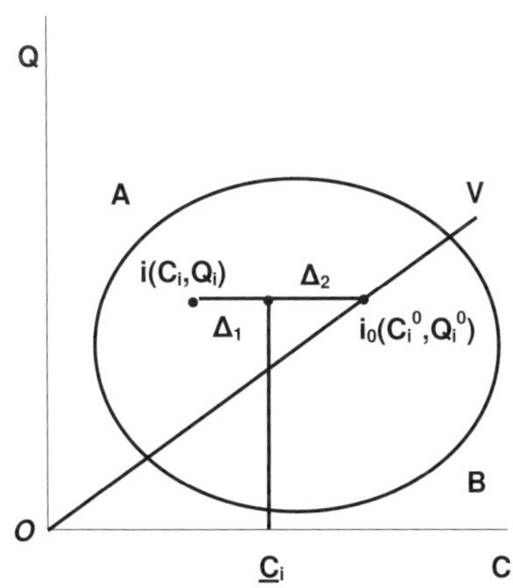

Fig. 1. Phase portrait of the behavior of market agents. Δ_1 and Δ_2 are the rent shares of seller and buyer, respectively.

The prime cost of a product unit C_i consists of labor cost for the producer i, including the labor pay and the living of the producer and his family. It is assumed that the producer sells his produce himself, without middlemen. If he is not very active, he sells the greatest part of his product, after a price fall, for a price close to the highest price Z_0. However, if a seller competes actively in the market with other sellers of high-quality goods, he tries to improve his marketing by charging a lower price whereby the price loss is cancelled by gain in sales volume.

In this chapter we study marketing dynamics meaning such an active seller and using a number of simplifying assumptions. First we consider the case of fair trade (competition) that fits the constraints of free market formulated in Chapter 9 and then the case when the constraint of fair quality information breaks down.

Fair competition

We consider the marketing dynamics of necessaries which are in use for a time much shorter than a human life and whose repeated purchase is controlled by human physiology or culture, such as a loaf of bread, a can of soda, a pack of milk, a tube of toothpaste, etc. These commodities are commonly very cheap and bad choices are a small matter, as people can easily move from one seller to another. Assume, again for simplicity, that the seller i charges a fixed price and is able to provide an invariable quality of his product, and can satisfy any great demand with sales increase. The assumption of invariably high quality can fit the conditions discussed in Chapter 9 only supposing that the i-th seller has a quite high efficiency P_i (see Fig. 5 in Chapter 9). As for the price, the actual practice is that it is more flexible being controlled by demand, but our simple model neglects these details.

Finally, we assume that the seller follows the free market constraints and provides reliable quality information when advertising his goods (Constraint 5, see Chapter 9). Such advertisement, rare in our times, is called fair advertisement. It is, for instance, a case of producers who declare the chemical composition of their products but can keep back application details. If the seller charges the highest market price C_i^0, he receives the total rent $C_i^0 - C_i$, or the maximum net profit, per unit quality (see the segment ii_0 in Fig. 1) but sales per unit time can be low; if he sells his produce for the prime cost C_i, his profit becomes vanishing. Therefore, he will try to reach an optimum gain (total rent of all sales per unit time) by charging some intermediate price $\underline{C}_i$ between C_i and C_i^0 (Fig. 1)

$$C_i < \underline{C}_i < C_i^0$$

at which the commodities would have a good sale and the rent taken for the seller's own would be rather high.

First we assume that advertisement is either forbidden or has no influence on buyers and these rely only on one another's opinions (as it used to be before the existence of mass media). Let the public know the fixed highest price for a piece of a commodity C_i^o, say a loaf of bread, of certain quality. Then the seller i who puts out bread of the same quality but for a lower price $\underline{C}_i$ ($C_i^0 - \underline{C}_i > 0$) can expect favorable reports from the buyers and a greater sale on the following day. Let sales increase proportionally to $C_i^0 - \underline{C}_i$. As each happy buyer of cheaper bread will recommend it to the same number of people, the sales increase by the same factor another day after, etc.: if on some day the sales reach K, they reach $M = aK$ on the following day, where the factor a is above unity and presumably proportional to the price change $C_i^0 - \underline{C}_i$. Therefore, we can here speak about "reproduction of sales" similar to reproduction in animal populations (see Chapter 1). Thus the process is amenable to the phase portrait modeling. The process of sales reproduction is shown by phase curve 1 in Fig. 2 where K is the number of sales on the current day and M is the number of sales on the following day. Then, sales can start with a very small amount and the phase curve may start from the origin of coordinates O. For some time, sales increase as a geometrical series at a constant "reproduction rate", whereby the ratio M/K remains invariable and phase curve 1 is a straight line. The curve is above the bisector as $M/K > 1$. The greater the difference $C_i^0 - \underline{C}_i$ the faster the sales growth and the steeper the phase curve (see the initial segment of a steeper curve 1a in Fig. 2).

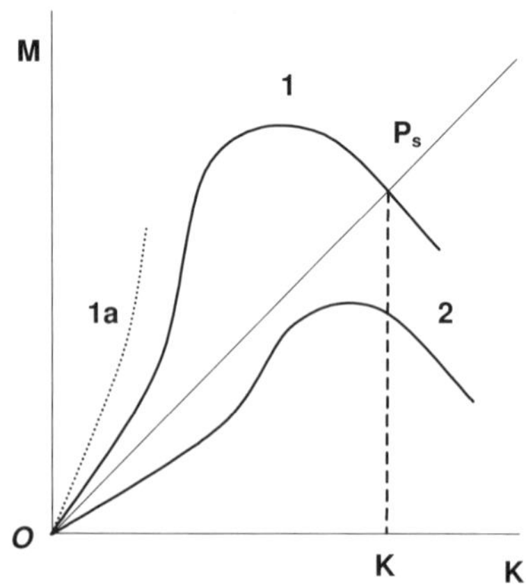

Fig. 2. Phase portrait of marketing dynamics (reproduction of sales). Curve 1 shows the sales of commodities in which buyers gain and curve 2 shows the sales of commodities in which buyers lose.

As soon as a quite large part of local population has been involved in buying bread of the species *i*, the process becomes controlled by a cooperative effect (related to the "herd instinct"). People learn about the cheap bread from many other people, not only from their friends, and the sales growth progresses faster than in a geometrical series; the M/K ratio grows, and the chord of the phase curve moves upward (Fig. 2). At such buildup of sales, the number of potential clients obviously becomes soon exhausted as the population is limited and each buyer needs a limited amount of bread for his family. Thus, the demand for bread stabilizes at some time: "reproduction" slows down, the slope of the chord OP to the K-axis becomes shallower, and the phase curve goes down to cross the bisector at some point P_s and on below the bisector (see Chapter 1). The number of everyday sales becomes fixed at the equilibrium point P_s at some distance along the K-axis (see the device of reflection from bisector in Chapter 1).

If $C_i^0 - \underline{C}_i < 0$, buyers think themselves losing (relative to the "standard" market price C_i^0 of bread of the same quality Q_i), avoid buying this bread and warn others against it. The phase curve of this process shows sales decrease ($M < K$) and moves down below the bisector. In population dynamics this shape means extinction and in our case the number of sales decreases to zero. This is the mechanism how too expensive goods are pushed away from the market.

On the other hand, commodities like the *i*-th quality bread preferred by buyers for its cheaper price also push out other producers from the "good" domain *A* (see Fig. 1)[a]. The producers along the line *V* who have the highest prime cost are obviously the first to quit. Their goods have no market and they move to the domain *B* of low-efficiency production. The line *V* rotates counter-clockwise and its slope shallows (Fig. 1) which means a decrease in the $Z_0 = C_0/Q_0$ ratio (market price of unit quality).

Yet, even active producers like *i* can get below the rising line *V* and thus become pushed away from the market by their more successful competitors. To avoid this they have to move left or upwards (Fig. 1), i.e., reduce the prime cost or improve the quality of their product. A successful competitor, who originally can have been a craftsman working alone or with his family, capitalizes on and hires labor men thus becoming a capitalist. This allows him to expand his production and introduce advanced technologies as he can afford paying inventors.

These laws of free market make commodities cheaper and of better quality which is rightfully interpreted as an important historic merit of capitalism

Unfair competition

Yet, the merits of capitalism are brought down by its faults (see Chapter 11). The appearance of mass media — newspapers, journals, and especially radio and TV — stimulated the progress of publicity which is

---

[a] The situation when the seller *i* has to increase production to meet the growing demand for his product and pushes out other producers from the market appears to be at odds with the theory in Chapter 9 which promised "safety" to all producers in the "good" domain. The difference is that the theory in Chapter 9 represents a "static" case while this chapter deals with the dynamics of marketing process associated with a growing producing potential of sellers.

their main source to live on. Even fair advertisement which offers reliable information about commodities and their applications biases the marketing process. These are not the best or cheapest commodities that benefit from publicity but the things whose producer is rich enough to pay the explicit and implicit potential of advertising. A minute of broadcasting time costs thousands dollars and the way and order of presentation are controlled by people who, in turn, depend on the advertisers. Popular sportsmen and stars are likewise employed for advertising various products. Thus the sales of advertised products are controlled by the money and PR capacities of the producers rather than quality. It appears even disputable whether "good" advertisement makes for fair competition as, anyway, it bars the way to better and cheaper commodities that lack such support.[b]

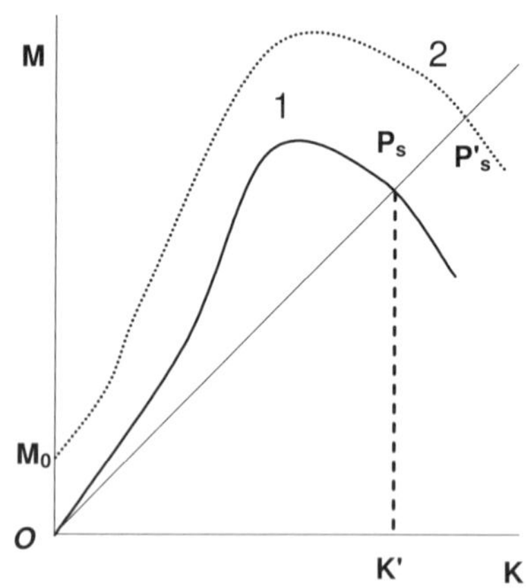

Fig. 3. Phase portrait of marketing dynamics under the effect of advertisement for high-quality and cheap commodities.

The effect of publicity can be illustrated using the phase portrait approach (Fig. 3). Curve 1 in Fig. 3 shows the sales of high-quality and cheap commodities ($C_i^0 - \underline{C}_i > 0$) not supported by advertisement and curve 2 shows the sales of the same products preceded and accompanied

[b] The question whether the publicity phenomenon agrees with the concept of free market is now of purely academic interest, and we leave it aside.

by advertisement. The pre-advertised commodities are demanded at their very appearance and the sales M_0 can be high already on the first day (Fig. 3).

Curve 1 is for sales without preventive advertisement and curve 2 is for sales preceded by advertisement.

Then favorable reports from happy buyers pushing up sales "reproduction" act parallel with advertisement which strengthens the cooperative effect and moves the sales curve up to the dashed line (Fig. 3). The dashed line crosses the bisector at the equilibrium point P'_s which is above the previous equilibrium point P_s. Thus advertisement provides a good start and, moreover, increases the stable number of sales.

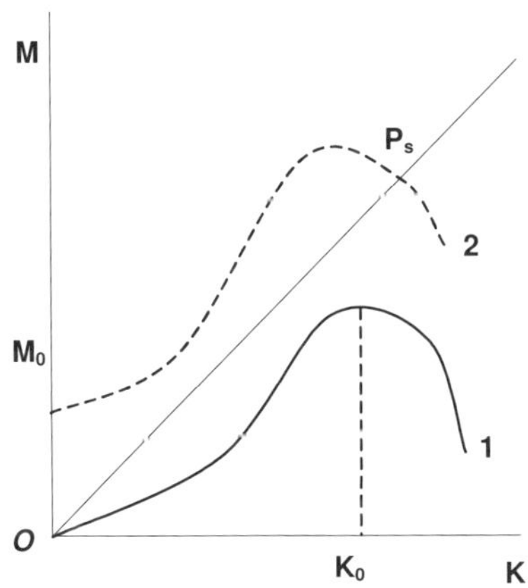

Fig. 4. Phase portrait of marketing dynamics under the effect of advertisement for low-quality and expensive commodities. Curve 1 is for sales without preventive advertisement and curve 2 is for sales preceded by advertisement.

However, publicity also rises the phase curve for too expensive commodities with their price above the standard market price or for low-quality goods, which otherwise would never have market (domain *B* in Fig. 1). Figure 4 illustrates marketing bias by unfair advertisement. Curve 1 in Fig. 4, like curve 2 in Fig. 2, traces the way of bad commodities pushed away from the market in the absence of advertisement: the curve is below the bisector already at the beginning, i.e., the "tomorrow" sales are lower than any today's sales, $M(K) < K$. In this case the amount of offered products K_0 should be high enough at the beginning of trade as no sales develop otherwise; some time later, the point (K, M) approaches O, i.e., the sales stop. On the contrary, the curve for pre-advertised products starts with the positive number of sales M_0 (Fig. 4) and then the cooperative effect encourages sales growth in the same way as unfavorable reports (negative cooperative effect) reduce the sales in the absence of advertisement. Finally, curve 2 crosses the bisector as demand cannot grow infinitely whichever be the advertisement, and equilibrium is inevitably achieved, following the common rule, at the point P_s. It means that a commodity which is in no way appropriate becomes imposed on consumers and comes into use as a customary thing.

In the case of a population of insects, curve 2 would correspond to manmade introduction of pests strange to the locality but bred, for purpose, elsewhere. Low-quality goods are just the same pests, where the laboratories of the producing firm act as breeding pools and advertisement serves for "import".

Bad commodities penetrate into market as soon as the cooperative or "herd" effect of public opinion overcomes the sense and personal opinion of individuals. The global effect of publicity is comparable to fascist or communist propaganda. Unfair advertisement keeps back the details of structure or composition of commodities but puts into the public mind ideas of their imaginary advantages using generalized statements. For instance, mass media advertise a useless drink known to be bad, which the constituent chemicals make taste disgusting. However, when hearing an advertisement claiming that "all America drinks stupid-cola" or seeing other people drinking it, an individual who rarely relies upon his or her own opinion believes that the "stupid-cola" does contain something good in it which he or she cannot appreciate. So people buy it again and again, more so it soon becomes difficult to find something better on sale.

The advertisement claiming that "all he-men wear stupid-something" or that "stupid-something makes every woman pretty", etc., does not need any reasonable background. The designers of advertisement purposely avoid anything capable of raising people's rational thinking. Publicity drives to closing of free market. "Conservatives" who expect the market to effectively solve today's problems should start with prohibiting at least some kinds of advertisement. Yet nothing happens, moreover, policy makers use the same PR methods to market their ideological products. The sway of publicity in the Western world has already undermined the natural trade mechanisms and flooded the market with low-quality expensive goods.

Finally, the last but not least: By suppressing the ability of people to pursue their own opinions advertisement clears the way for totalitarian choices which mask themselves under an appearance of democracy.

Chapter 11

Labor Market and Capitalism

Labor market

In classical economics labor is a measure of the work done by people and is a key factor of production. The view of labor as a commodity is actually very old — hired labor existed already in ancient times — but the market of labor appeared with capitalism, which differs the latter from any other mode of production. It was David Ricardo who introduced the concept of labor as a commodity to economic science. Yet labor is a special commodity. In the context of free market relations (see Chapter 9), each wage worker should be a seller of his labor and this labor should have its productivity and labor cost. Thus we discuss labor as a special commodity under capitalism and view capitalism as it exists nowadays rather than in the time when it began and was not yet the dominant economy.

In the simplest case we leave aside the quality of labor and thus assume that the efficiency of every wage worker who sells his labor of some type (turner or mechanic, or any) is measured by the amount of product he can manufacture per unit time, say, per hour. For numerical evaluation of efficiency, labor in a given market is assumed to be uniform in type and quality and measured in certain units, say, in pounds or in pieces of product. Or, rather we say that the efficiency of a laborer is P units. Actually it makes no difference which units to use because we consider a market of labor as a special commodity and estimate labor productivity in manufacturing the same things.

Labor cost is a more difficult point. Marx tried to apply the general Ricardo's approach and measure the value of labor, as in any commodity, against average abstract labor-time socially necessary for its production. By production of labor power Marx meant the required input into its "reproduction", i.e., bringing up children, gaining skills, etc. Our definition of labor cost is independent of the value theories but stems from empirical data on labor market.

When we discussed growing grain in Chapter 9, we evaluated its prime cost via the minimum hourly pay: the labor cost of the i-th producer (S_i) was assumed to equal the pay he would get for the same labor time as if he were hired as a work-hand. This assumption was used to explain why nobody sells his produce below its prime cost. Now we generalize this approach and assume that the i-th laborer, with the known cost of his living (subsistence) and working skills, can get an hourly pay at least equal to S_i if he offers his services at any labor market[a] of a given place in a give time. Unlike the case of Chapter 9, S_i, does not include expenses for maintaining the means of production as a wage worker has no this possession. On the other hand, the minimum pay has already determined his living standard — ambitions and habits of himself and his family — which is defined by specific local labor conditions. The labor conditions, in turn, depend on laborer's skills in a general sense of ability to live on any labor, as well as on social factors like sex, age or race all employers always take into account. Nevertheless, S_i is independent of any other factors but is the minimum the i-th laborer always gets whichever be the work he does. He knows this value, as well as his efficiency P_i and thus knows the prime cost of his labor $C_i = S_i/P_i$ (or, simply, "his prime cost").

No laborer obviously accepts to work for pay (wages) below his prime cost. When being hired, he declares his efficiency P_i (most often as a skill grade) but does not declare his labor cost S_i to the employer, the only buyer of his labor. Efficiency is easy to check in the course of work, and as for labor cost, the employer can predict it himself and take into account. Labor cost is a value which shows up at the first stage of labor market (corresponding to the first season of grain market): at this stage wages reduce to the prime cost as a result of competition between

[a] In Chapter 9 we dealt with a specific case when a producer can only grow grain.

laborers, like in any market, and the further market stage develops at a fixed labor price.

Thus, each laborer i is represented by two parameters S_i and P_i, and, correspondingly, by the point (S_i, P_i) in the plane (S, P). Then, we can plot a cloud of points, or the phase portrait, in the same way as in Chapter 9 (Fig. 1):

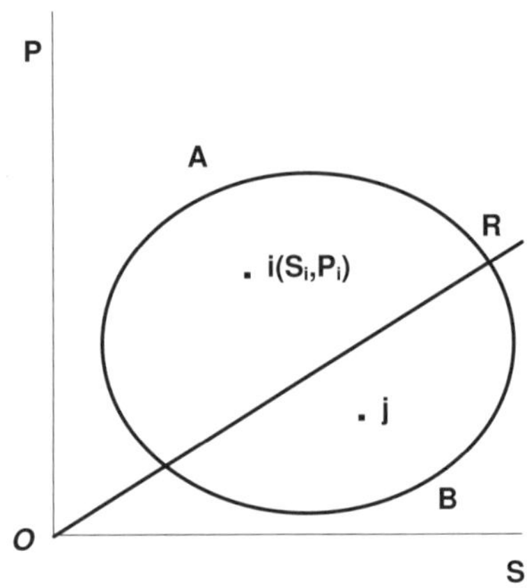

Fig. 1. Phase portrait of free labor market in the plane (S, P).

The points $i = 1,2,\ldots, n$ show n laborers that offer their labor in the market, and the straight line R divides the cloud into the domains A and B. Modeling labor in terms of the parameters P and S imparts to it the principal features of a commodity. As in the case of Chapter 9, we assume free market in which the market agents behave according to the same rules but the capitalist, the owner of an enterprise, is the only buyer and the laborers are sellers.

The total hourly output of all n laborers is

$$P = P_1 + P_2 + \ldots + P_n$$

units (pounds, pieces, etc.). Let $\underline{P}$ be the total demand for the commodity the laborers manufacture per hour. Then, if the market fully satisfies this demand, the value $\underline{P}$ measures the total demand for labor of a given type, as labor is measured in the same product units. As before, we assume that

$$P > \underline{P},$$

that is, the absence of labor shortage. Then the wage workers who get jobs at the given labor market are shown as the points i in the domain A above and along the line R, and the other laborers j, belong to the domain B and are unemployed (Fig. 1). The existence of unemployment satisfies the free market condition of no labor shortage.

Let the number of wage workers be m ($m < n$): 1, 2, … , m, like in the case of Chapter 9. Then, by definition of the domain A where the respective points belong,

$$\underline{P} = P_1 + P_2 + \ldots + P_m,$$

i.e., m laborers fully provide the required output. On the other hand, the total labor cost in the given labor market is

$$S = S_1 + S_2 + \ldots + S_m,$$

which is the least possible value at the above total output, as it was proved in Chapter 9 and holds for any free market. Again, free market solves this optimization problem without any special planning just by spontaneous market behavior of wage workers, the sellers of their labor, and the capitalist, its buyer.

Wages

The i-th laborer has his labor at the prime cost

$$C_i = \frac{S_i}{P_i},$$

like the prime cost of any commodity. The fixed price of a labor unit C_0 is found in the same way as in Chapter 9 and is the highest prime cost C_i in the domain A; for all points along the straight line R,

$$C_0 = \frac{S^0}{P^0}$$

(Fig. 2). Figure 2 shows that $C \leq C_0$, and the equality occurs only along R.

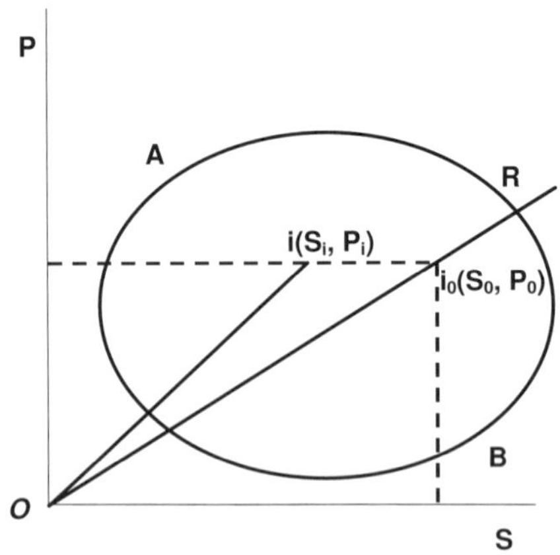

Fig. 2. Phase portrait of wages.

The hourly labor of the i-th laborer (P_i) is sold at the fixed price C_0, and the laborer's hourly pay is thus

$$S_i^0 = C_0 P_i .$$

We can express this value geometrically as follows. If we draw a horizontal line through the point i until it crosses the line R at the point i_0 (Fig. 2), S_i^0 is the S-coordinate of the point i_0. Obviously, $S_i^0 \geq S_i$, and the equality is possible only for the points i that belong to R. The difference

$$DS_i = S_i^0 - S_i$$

can be interpreted as rent got by the i-th laborer. If S_i corresponds to his habitual living standard, rent is the excess increasing at increasing efficiency P_i.

The total input to wages at a given enterprise is

$$S^0 = S_1^0 + S_2^0 + \ldots + S_m^0 = C_0(P_1 + P_2 + \ldots + P_m) = C_0\underline{P},$$

while the average pay for a product unit equals the labor price C_0:

$$\frac{S^0}{\underline{P}} = C_0.$$

The capitalist obviously follows a constant incentive to reduce this value by decreasing the S_i/P_i ratio for his laborers, which corresponds to lower mean wages $C_0 = S^0/\underline{P}$. The S_i/P_i ratio can decrease either by increasing the labor productivity P_i or by saving the cost S_i; reducing S_i^0 is tantamount (Fig. 2) because P_i enters the denominator $\underline{P}$ and S_i^0 enters the numerator S^0 of the C_0 equation.

The early "wild" capitalism did not distinguish labor from other commodities. People who could not find jobs faced starvation and perished, just like unsold goods. The famous Victorian age, which is believed to be the acme of the Western civilization, rested upon the poverty of laborers called the working class, or labor, just at that time. Market, then free from any governmental regulation, led to unrestrained competition among capitalists who strove for reducing the price of their goods by saving on wages. Although machines were already in use, they were imperfect, and much hand work was required anyway. Therefore, extending the working day was the most common way to enhance labor productivity. At the same time, capitalists looked for the least ambitious laborers, i.e., those with the lowest labor cost. In the Victorian time they were women and children. Under early capitalism, both possibilities to reduce the S_i/P_i ratio led to the limits that appear quite incredible nowadays. The working day could reach as long as sixteen to seventeen hours, usually without days off. The living standard of workers fell below the subsistence, which made them send their wives and children to factories, contrary to the English ways. Children started working since the age of five or six, as soon as they grew able to figure out what they were supposed to do. That practice, called *exploitation*, outraged the laborers themselves and the public.[b] Marx extrapolated the tendency of

[b] Children's labor was especially outrageous and was first blamed by medical people, priests, and writers, most often people of moderate views. Their protests called forth parliamentary investigations and, as a result, a law was accepted in 1819 which limited the working day of children to twelve (!) hours. Less moderate opponents against the unfair social ways came out later: the term 'socialism' first appeared in 1832 and the

pay saving to the future and arrived at his hypothesis of the "absolute impoverishment of the working class under capitalism", which we discuss below.

Revolts of wage workers forced the government to interfere with market. The Luddite movement of frame-breakers in England, revolts of Lyons weavers in France and Silesian weavers in Germany eventually made the European governments restrain the market freedom, because the very market edifice was prone to collapse and hit the moneymakers. The English government was the wisest in this respect, though the parliament reform of 1832 and the ensuing appeasing measures were undertaken only for the threat of pending revolution in the early 1830s. The continental governments chose coercion which brought to three revolutions in France driven by working people (1830, 1848, 1870) and, eventually, to the same legal regulations as in England. The social unrest in Germany was redirected towards nationalism but the final result was the same — after two world wars. Nowadays a free market of the Victorian-age kind exists nowhere in the world. Some economists keep arguing that any intervention into market offends against civil freedom and brings about other offences. However, free market has been restrained in all developed economies and the restrictions thus cannot be attributed uniquely to "bad ideas". A perfectly free market is apparently as unfeasible as any perfect institution. The problem is where to draw the line not to deprive market of its essential advantages.

A control measure that holds back the fall of the living standard in the lowest paid wage workers was invented in the twentieth century: it was the law of floor wage that forbid wages below some amount S_{min}. The consequences of this law for the labor market are illustrated in Fig. 3. First, the wage workers with the S-coordinate below S_{min} become excluded from the domain A, i.e., those who belong to the part of A between the line $P = W$ and the line R lose jobs. As a result, the total output $\underline{P}$ decreases and the loss has to be compensated by hiring some workers who were formerly unemployed, or belonged to the area below R. Geometrically it corresponds to the clockwise rotation of the line R to R'. This makes the domain A' above R' wherefrom we have to cut off

---

word 'communism' acquired its modern meaning in 1840.

workers with wages below S_{min} who belong to the area between the lines $P = W'$ and R' (shown by horizontal hatching). According to the figure, some of the formerly dismissed workers regain their jobs and the formerly unemployed that belong to the area between R and R' above the line $P = W'$ can obtain jobs as well. The output of the added areas has to keep up the missed output of the cut-off area (horizontal hatching).

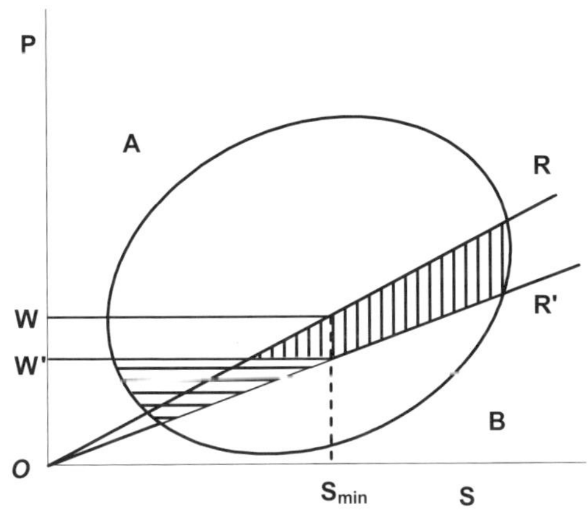

Fig. 3. Dynamics of labor market associated with the law of floor wages

Nowadays laws that guarantee the minimum wages work in all large economies. Note that at the normal form of the labor power cloud, the clockwise rotation of the line R means that the mean wages increase (Fig. 3), i.e., higher-paid laborers substitute for lower-paid ones. This is a reason, not the most important one, why Marx's principle of absolute impoverishment fails.

People that lost their jobs with the establishment of the floor wage limit, like other unemployed, enjoy a dole for subsistence and apartment rental and free health care. Unemployed, if they are sound, are supposed to learn another trade or improve their efficiency.

The system of reward to unemployed was put into practice for social rather than economic reasons: it was an attempt of the authorities to cope with social troubles, especially among Afro-Americans, paying off with sops. It was reasonably criticized as giving rise to parasitic attitudes and unreasonable expansion of administrative offices. However, the partisans of the welfare system stress its economic benefits, as the rewarded unemployed remain consumers and buy goods that would be inaccessible for them otherwise, which maintains the level of production and gives jobs to others. In our view, the latter ideas can hardly excuse the existence of a permanent group of unemployed, sound people in their full active age, who lose interest in work. We give more attention to this problem in a special chapter below (Chapter 12).

Capitalistic production

In our discussion of labor market so far we have meant a single enterprise and its specific product. Now we take several enterprises that compete in manufacturing the same commodity. We apply the term *capitalist* to each decision-making owner of such enterprise, though they are rather a group of people than an individual nowadays. Let capitalists be denoted as i and S_i be the total yearly expenses of the i-th capitalist on maintaining production (tools, raw materials, power, and labor pay). Let M be some annual amount of product, conventionally called "mass of product" which can stand for its volume, weight, or pieces, or any unit. First we consider the case of invariable quality. Let the i-th capitalist be represented by his yearly input S_i and yearly output M_i. Then, we can show capitalists as the points (S_i, M_i) in the plane (S, M) (Fig. 4) in the same way as grain producers in Fig. 1 of Chapter 9. All these points make up a cloud of n capitalists who can put out a total of

$$M = M_1 + M_2 + \ldots + M_n \,.$$

Assume that M exceeds the demanded quantity $\underline{M}$:

$$M > \underline{M}.$$

Thus, it is again a shortage-free market, as any free market usually becomes.

We draw the straight line R, in the same way as for the grain market, to cut off the domain A including the points i (above and along R) that yield exactly the demanded quantity $\underline{M}$ and are numbered as 1,2,..., m ($m < n$). Then,

$$\underline{M} = M_1 + M_2 + \ldots + M_m.$$

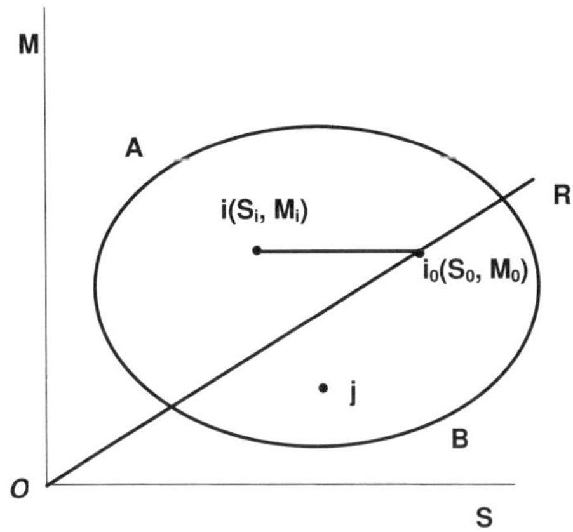

Fig. 4. Phase portrait of competition.

The remaining points j make up the domain B and are out of production. The respective capitalists should ruin and leave the market, if it is really free[c]. Capitalists that belong to the domain A supply the demanded amount of commodity $\underline{M}$ at the least total costs:

$$S = S_1 + S_2 + \ldots + S_m.$$

This is the result the appropriate optimal planning should reach for the community benefit. Yet, this planning is impracticable for the lack of exact data on all producers and consumers (this knowledge is presumably available only for a hypothetical scientist who performs phase portrait modeling as in Fig. 4) and for the unfeasibility of exhaustive calculations. Like in all other cases, the problem is in the complexity of the system. The economic theory can serve as a clue to understand the ongoing processes rather than as guidelines to undertake

---

[c] Governments in modern economies as a rule help these companies for some reasons, but support to capitalists is casual being not legally secured. Therefore, we leave this case beyond our consideration.

a priori planning of production and consumption. That was the idea in the command economies, called "socialistic", where people tried to meet overwhelming challenges and failed. Meanwhile, free (or at least partly free) market is capable of equilibrating things through competition in which each producer and each consumer just pursue their own ends. The explanation of this process in Chapter 9 is general and holds for all market phenomena.

The prime cost of a unit product for the i-th capitalist is

$$C_i = \frac{S_i}{M_i},$$

and the highest prime cost C_0, reached by the points i_0 along the line R (Fig. 4), is

$$C_0 = \frac{S_0}{M_0}$$

The value C_0, according to the general law of free market (see Rule 1 in Chapter 9), becomes the fixed price of a unit product. The S-coordinate S_0 of the point i_0 is the annual revenue of the i-th capitalist who sold his yearly output product for the fixed price, and the S-coordinate S_i of the point i is his yearly input into production. The difference corresponding to the length of the interval ii_0 is the rent:

$$DS_i = S_0 - S_i.$$

The analogy between capitalistic production and growing grain ends at this point. Growing grain is a particular case of the system we describe in this section and it only approximately approaches capitalistic production. The distinctive feature of the latter is that capitalist can invest his rent into expansion of production, i.e., increasing the output M_i. A landowner using a piece of land likewise can increase its rent by improving the land quality (due to fertilizers, etc.) but, unlike capitalist, he has limited possibilities. Indeed, increasing the crop capacity becomes uneconomic at some point requiring expenses that exceed the rent growth. This regularity is related to the natural properties of soil, the indispensable mean of production in cropping. The advanced ways of land use moved the fertility limit quite far ahead but the limitation remains because food production cannot go without photosynthesis and thus depends on natural conditions (see more to this problem in Chapter 6). However, products of industry are amenable to unlimited upgrading.

Expansion of production

A capitalist can spend a part of his rent either for simple production growth or for technology improvement, most often using innovations bought from some inventor. Both cases provide a larger output but require a greater input. We assume for simplicity that the output M is proportional to input, as it occurs roughly this way in simple production expansion. Then the coordinates S_i, M_i of the point i increase by an equal factor, say, k-fold, i.e., the point i moves along the ray Oi to the position i_1 for the corresponding distance away from the origin (Fig. 5). This motion changes the geometry of the "cloud of capitalists".

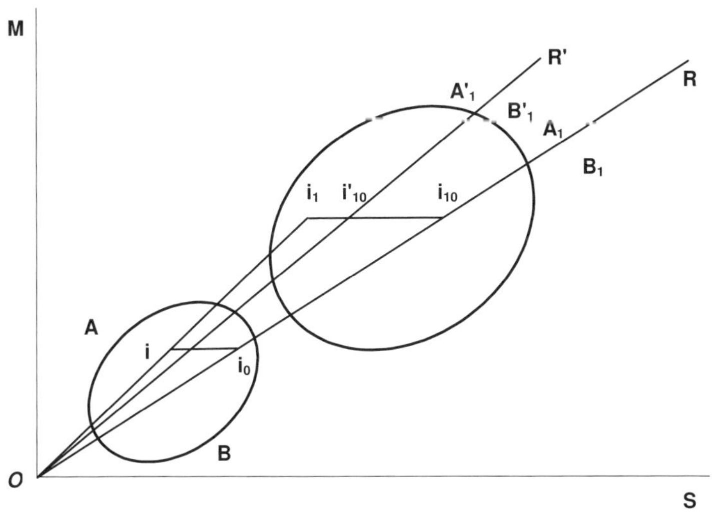

Fig. 5. Phase portrait of production expansion associated with technology advance, at $k = 2$.

If only one capitalist increases his output supplying a minor portion of the demanded amount $\underline{M}$, neither the cloud changes much nor the position of the line R. The k-fold increase in rent is evident in Fig. 5 (interval ii_0), under the assumption that the line R is fixed and, correspondingly, the commodity price is fixed as well. This is the advantage of any innovator, provided his innovation is pioneering.

However, things take a different turn if all other capitalists decide to expand their production, for instance, having learnt the know-how of the capitalist i. Then, a capitalist formerly represented by the point i with the coordinates (S_i, M_i) increases his output k-fold and, according to our assumption, increases his input correspondingly, to arrive at the point i_1 with the coordinates (kS_i, kM_i) (Fig. 5). Thereby the cloud of capitalists transforms into a new cloud by k-fold expansion of the plane (S, M). This is a similarity transformation in which every point i moves along the ray Oi away from the origin. The expansion transforms the line R into itself and the domains A and B into A_1 and B_1, respectively. The domain A_1 includes the same number of points as A but the total output increases k-fold. To fit to the invariable total demand $\underline{M}$, the line R' cuts off some part of the domain A_1 (let it be A_1') with the k times lower output (Fig. 5). This procedure equilibrates the output with the total demand $\underline{M}$ but reduces the rent of each capitalist who expended his production so that the rent becomes $i_1 i'_{10}$ instead of $i_1 i_{10}$ (Fig. 5). Moreover, the point i_{10} can fall below the line R' and the respective innovator can go broke. Indeed, any technology breakthrough inevitably drives some part of producers out of business.

If all capitalists expand their production k-fold, the new fixed commodity price C_0' measured by the slope of the straight line R' to the M-axis is lower than the previous price C_0, i.e., the commodity becomes cheaper.

Now consider the impact of technology advance on wages. Turn back to Fig. 2 and assume that the efficiency of each laborer increases k-fold due to some technological innovation. If the total demand remains invariable, and the labor productivity is measured by output as before, the total labor demand $\underline{P}$ is invariable as well. On the other hand, the labor costs of each worker S_i must likewise remain invariable if these are the same laborers with the same living standard. In this respect, labor is a special commodity as a wage worker, its seller, increases his efficiency without increasing his input into the commodity he sells (his labor): it is usually the capitalist who takes on expenses for technology advance[d]. Therefore, the laborers formerly represented by the points i with the coordinates (S_i, P_i) move to the points i_1 with the coordinates (S_i, kP_i) as their efficiency increases k-fold due to the technology advance while the labor input remains invariable (Fig. 6).

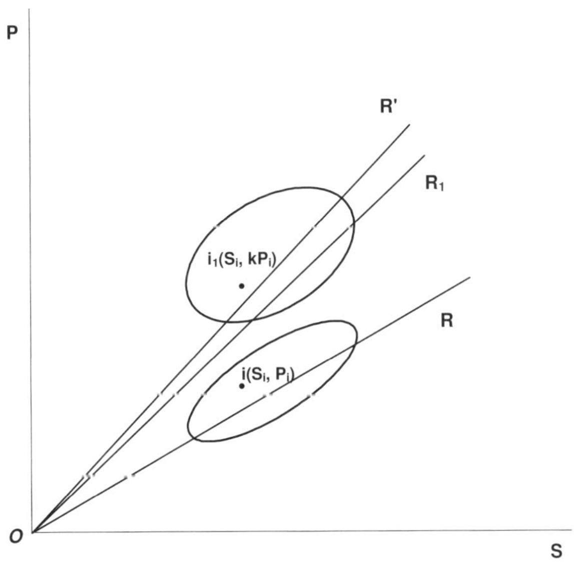

Fig. 6. Phase portrait of wages associated with technology advance.

The cloud of the points i transforms into the cloud of i_1 by the k-fold expansion of the plane (S, P) upward along the P axis. This transformation brings the line R to R_1 and the domain A to A_1 cut-off by the line R_1. The slope of R_1 to the P-axis is obviously k-fold lower than the slope of R. The points i_1 above and along the line R_1 exactly correspond to the former points i which were above and along R, i.e., the

[d] We leave aside the case when laborers have to grade up their skills because the capitalist most often pays for their training as well.

domain A_1 includes the same laborers who formerly belonged to A but their efficiency increased k-fold. Therefore, they yield the amount of labor $k\underline{P}$ whereas the labor demand $\underline{P}$ remains the same, as the population is assumed invariable and consuming the same amount of commodities. Of course, the capitalist gets rid of the excess laborers and leaves as many as is needed for their labor to fit the amount $\underline{P}$. Then, the cloud A_1 becomes truncated to its part A' by some line R' above R_1 (Fig. 6). The line R' determines the labor price for those who kept their jobs after innovation. The line R_1 being steeper than R_1, its slope to the P-axis is lower, i.e., less than one k-th of the slope of the original line R.

Inasmuch as it is the slope of the line R' that defines the labor price after innovation, C_0 decreases correspondingly at least k-fold. However, the laborers who kept their jobs have k-fold greater output, and their wages C_0P_i decrease (at this point we do not evaluate the decrease). On the other hand, we showed above that technology advance causes price fall for all commodities except labor. A more thorough investigation shows that the decrease in wages is of the same order as the average price fall but higher than the latter.

This inference supports Marx's principle of absolute impoverishment of the working class, which did work at some evolution stages of the capitalistic society. Note, however, that our model included the technology advance meant as growing output of the same commodities and neglected quality and stock variations, in the same way as Marx did. The actual consequences of technology breakthroughs are much more complicated than that — as we show below — and mean a new stage in the history of capitalism.

Quality and technology advance

Technology advance increases labor productivity and, what is more important, provides quality improvement. Mere production expansion typical of early capitalism implies expansion of market or onset of new markets as it was the case of England when hand weaving gave way to factory manufacturing of textiles which were exported to India and Indian weavers ruined, or the case of America and other British colonies which required more metal works. At that time the idea of progress in technology was associated with increasing production of the same "standard" commodities. Yet, quantitative production growth has its natural limits: under capitalism it is ahead of population growth. The demand for necessaries becomes thus fully satisfied and people care more about the quality of goods. This has occurred in large economies where even low-paid population groups do not accept to eat and wear what their grandfathers ate and wore and do not want old houses, at least since the last century. Therefore, capitalists have to improve quality to make goods marketable.

Another limiting factor, especially notable nowadays, is the Earth's natural resources, the fact which was long neglected before. The ever growing use of some commodities threatens the environment. As we wrote in Chapter 4, the technological burden can become unbearable for the environment if the third world follows the large economies in the use of cars and other vehicles.

A special emphasis on quality is the basic feature of the today's technology. In the model that accounted for quality in Chapter 9 we used two parameters (prime cost $C = S/P$ and quality Q) because including three basic parameters (labor cost S, output P, and quality Q) would require a sophisticated mathematical treatment. With this restriction, each capitalist is represented by the point i with the coordinates (C_i, Q_i) in the plane (C, Q) where C is the prime cost of a commodity (say, in dollars per pound) and Q is its quality (per unit product, say, per pound). Figure 7 images the respective cloud of the points i in the same way as in Fig. 8 of Chapter 9.

Further we assume that the output P_i is known for all capitalists i. Then, if the total output of all capitalists of the cloud is

$$P = P_1 + P_2 + \ldots + P_n,$$

and in the absence of shortage, the total demand $\underline{P}$ is below P; we can draw the straight line V in Fig. 7 such that the capitalists (1, 2, ..., m) with the total output $\underline{P}$ were above and along this line:

$$\underline{P} = P_1 + P_2 + \ldots + P_m.$$

The capitalists corresponding to the points 1,2, ..., m occupy the domain A and their enterprises are running, while the others, which belong to B, are bankrupt.

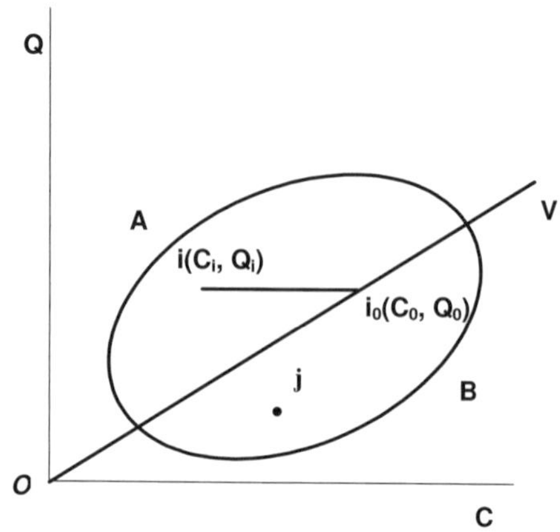

Fig. 7. Phase portrait of production with regard to quality in the coordinates C (prime cost) and Q (quality). Minimization of price for a quality unit.

In the process of pricing (Chapter 9) buyers strive for reaching the best quality commodities of those available in the market. Or, more exactly, buyers choose the seller with his goods of the highest possible quality per dollar.[e] The fixed price for a quality unit is

$$Z_0 = \frac{C_0}{Q_0},$$

where the ratio in the right-hand side, the prime cost per quality unit, is the highest along the line V. The domain A (Fig. 7) is the solution to minimization of price for a quality unit at a given total output. The rent of the i-th capitalist, i.e., his profit from quality, is

$$DC_i = C_0 - C_i$$

and is measured in dollars per unit product (say, per pound). Geometrically the rent corresponds to the interval ii_0 (Fig. 7).

---

[e] Note again that we mean quality equivalent to von Neumann's utility, which is proportional to quality.

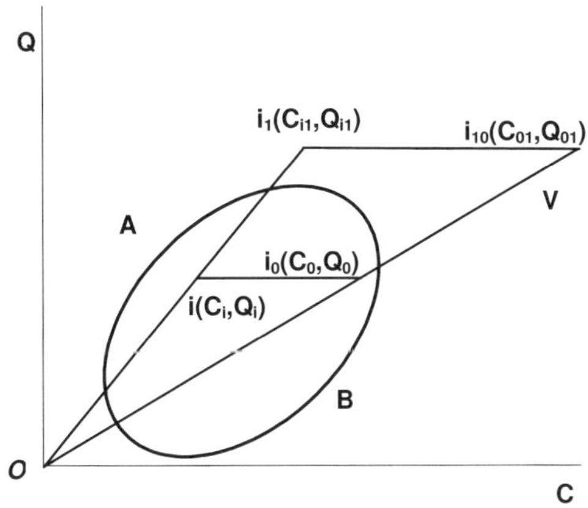

Fig. 8. Phase portrait of production dynamics under the effect of technology advance with regard to quality.

Now imagine that advance in technology allows a *k*-fold quality increase in all producers, and the prime cost proportional to quality increases correspondingly. Then, the capitalist *i* with the coordinates (C_i, Q_i) moves to the point i_1 (Fig. 8) with the coordinates $C_{i1} = kC_i$, $Q_{i1} = kQ_i$, and the domain *A* of running enterprises moves to the new domain A_1. All capitalists obviously increase their rent *k*-fold (Fig. 8). If the output *Pi* of all capitalists is assumed invariable[f], the producers that stay in the market remain the same and produce the same demanded quantity $\underline{P}$ of commodities. Therefore, the new domain A_1 is cut-off by the same line *V* and the price of the unit product Z_0 does not change. If the progress decreases the prime cost of a quality unit, i.e., reduces the labor cost per unit quality, all points i_1 in Fig. 8 move leftward and the line *V* thereby approaches the *Q*-axis; therefore, its C_0/Q_0 ratio decreases and each quality unit becomes cheaper.

If wages remain invariable at invariable output, population gets an opportunity to consume more quality units than before the technology advance. Thus, commodities acquire a greater utility in terms of von Neumann's theory in which utility measures the willingness to exchange one commodity for another. In a common sense, people enjoy higher-

[f] Mind that we are investigating only the impact of technology advance on quality.

quality commodities, as it occurs in developed economies and is often referred to as higher "quality of life". The term appears not very exact, however, as life by no means reduces to consumption, while developing technology not necessarily improves other aspects of life, for instance, the environment.

Note that technology-related quality increase without production growth leaves enterprises running (the line V is fixed) and does not aggravate unemployment. Finally, even if a decrease in prime cost rises the line V, i.e, some capitalists quit, the appearance of new types of production gives rise to new jobs. In terms of Chapter 9, there arise new planes (C, Q) for new commodities. Thus, the orientation of technology progress toward higher quality and diversity of production obviously increases the living standards of population.

Theories of value

It appears pertinent to dwell upon the concept of value in our discussion of capitalism. *Value*, not to confuse with market value or market price, is a special theoretical concept that originally belongs to David Ricardo. Ricardo believed that every commodity would bear some numerically expressed "natural" value which depends neither on the way it was manufactured nor on market. Ricardo, a prominent scientist, is responsible for developing theories of rent, wages, and profits but his concept of value became one of key errors in the history of science[g]. Marx who was Ricardo's follower in his many ideas extended the concept of value to labor power. Thus he discovered the profit or surplus-value which arises when workers do more labor than is necessary to pay the cost of hiring their labor-power, or, in other words, the value of commodities is higher than the value of labor a laborer puts in. Marx thought surplus value to disclose the secret of capitalistic exploitation. Thus he pretended to have theoretically underpinned the idea of

---

[g] He himself wrote: "...from no source do so many errors, and so much difference of opinion in that science proceed, as from the vague ideas which are attached to the word value" [*Ricardo*, 1821].

exploitation and explained a phenomenon which was merely a matter of outrage before. We leave aside the problem of exploitation and try to look into the very concept of value.

The concept of value was inspired by the scientific air of the time when it made its first appearance, the early nineteenth century. At that time mechanics, which already developed the concepts of mass and energy, was taken for a model exact science. Earlier Antoine Lavoisier suggested the mass conservation principle, and the principle of energy conservation in its mechanic form was well known though without the term "energy". The latter principle was already known in the general form and was proved valid by James P. Joule's experiments in the time when Marx was working on his *Capital* since the 1850s. The concept of energy is especially close to what Ricardo meant.

In mechanics each body is characterized numerically and the parameter of energy is closely related to the concept of work. Energy depends on the body's state: a moving body has a greater energy than a static body and a body above the Earth's surface has a greater energy that that on the surface. Transition from one state to another requires work and the value of this work, by definition, is assumed to be the energy increment associated with transition to a different state. The energy increment is given by

$$U(B) - U(A) = W(A, B),$$

where $W(A, B)$ is the work required for moving a body from the state A to the state B and $U(A)$ and $U(B)$ are the energies of the body in the state A and state B, respectively. This equation makes sense if work in its right-hand side is independent of the way of conversion but is absolutely defined by the initial and final states. The mechanic conditions when this equation is valid were known as early as in the end of the eighteenth century. The gravity and electrostatic forces satisfied those conditions and were used to express work. Furthermore, the equation of work gives the relative rather than absolute energy, and the latter can be evaluated if the energy in some initial state (say, the state A) is assumed zero; then the value of $U(B)$ can be found for any state B. The arbitrariness in the choice of the initial state means that it is the energy change that matters in mechanics rather than its absolute numerical value. What is important in the above definition of work is that work can be evaluated through a unique procedure as soon as the initial and final states of the body are specified.

Ricardo proceeded from the idea that making a commodity out of raw material or a half-finished product is equivalent to conversion from one state to another by means of work. He hypothesized that every commodity can be assigned some numerical characteristic that increases with the input work (labor) and called it "value". To that point the analogy with mechanics appears obvious: one has to estimate the work applied to make a finished product and prove it to be independent of the way how the product was transformed from its initial to final state. Ricardo tried to measure this work against labor time but faced difficulties no economist was ever able to overcome. Indeed, workers can have different skills or use different technologies or especially machines that can save much labor time. Thus, different work can drive to the same end. Being aware of that, Ricardo invoked "socially necessary" labor which refers to the quantity required to produce a commodity in a given state of society, under certain social average conditions or production, with a given social average intensity, and average skill of the labor employed [*Marx*, 1865].

In this understanding, value becomes dependent on many intricately related factors that elude unambiguous definition, and hence, unlike energy, value cannot be calculated. Ricardo's concept of value thus turned out to diverge with the concepts of theoretical mechanics known at that time where energy is measured in a one-dimensional scale. Ricardo's value is a scholastic construction which rather confuses than clears up economic issues. This scholasticism ruled out understanding of value based on experience and was often reasonably criticized.

Nature of capitalism

Capitalism can be defined as a mode of production based upon labor market. A detestable property of labor market is that man is taken for a commodity. Of course, a wage worker sells his labor for a while rather than himself, and the previous statement is not quite exact. Yet, a wage worker is fully dependent on the employer and his labor approaches slavery at the point that the product is as a rule strange to the worker and his personal feelings and preferences. The use of machinery reduced labor to a number of monotone movements. In this respect, machine labor is worse than the labor of a farm-hand who deals with nature and is fully aware of what he is doing. A peasant who cultivated his land or an artisan in his workshop were free in decision making and responsible for their business and related risks. This position better fits the human dignity than the position of a modern wage worker who would be easily replaced by a robot if the latter were as cheap. Perhaps, robots will save ever more human labor as far as they become cheaper.

The above ideas were often criticized meaning that any work would be tiring and monotone and work by itself would be humankind's perdition (punishment for the original sin if you will); all annoyance of wage labor would be paid off by the absence of personal responsibility and higher consumption standards. Furthermore, some argued that people would depend on one another in any society and the dependence of a peasant on a feudal or a handicraft on the imposed guild rules were by no means better than the dependence of a wage worker who can quit when he wills. Of course, these arguments sound reasonable as far as people's preferences are controlled by aversion against familiar things and illusions about the things that are beyond their personal experience. (Yet, illusions about the past are always doubtful and any attempt to turn back into the past fails.) However, the instrumentalistic idea of wage labor as an inescapable evil rewarded by the lack of responsibility and better consumption appears to stem from the pessimistic attitude of doom whereas the idea that people are not created to work as machines, which we share, at least presumes changes in the existing ways. Anyone who has ever worked at a plant or operated a machine just to earn his living knows that he never wanted to, and nobody ever wanted. Therefore, the sooner robots can replace people in this work the better, though there arises the problem how to employ those people who never tried any better alternative. On the other hand, the problem of exploitation will persist as long as wage work exists.

Moral indignation against exploitation of wage workers is independent of any theory of value and has no relation to any scientific proof. It rather roots in the inherited system of moral values that blame undeserved appropriation of rent and viewing man as a commodity. The existence of rent causes no doubt. The moral aspect is whether getting rent is rightful. Even though the capitalist invests a great portion of rent into production growth or upgrading, he takes the decision how much to hold for himself and how much to invest, while this decision not always makes for the enterprise success. Moreover, a today's capitalist himself is no longer the manager of production as the early capitalists used to be, and the management is given up to hired specialists. In a general case, capitalists get a considerable part of rent, hard to evaluate, which is due to the property ownership rather than to their own labor. Many people feel this unearned appropriation of rent unfair and call it exploitation. Of course, the judgement of whether this reward is earned or not depends on moral attitudes and is not a subject of economics. The vague idea of surplus value adds nothing to the point.

The capitalistic mode of production did eventually bring population to higher living standards (after ages of wild capitalism). Capitalistic production may require the existence of capitalists, and in this case their profit has no relation to their personal merits but is a sort of inevitable tax needed to maintain the "affluent society". Capitalists themselves would hardly like this "existential" justification as it gives them the part of parasites, though useful for the community. If this justification is wrong, it is reasonable to look into what the rent holders actually do for production.

Note that the today's capitalism departed far away from free market, not only due to intervention from the government but mostly as a result of monopolies and impudent publicity.

Another essential note concerns the relation of capitalism with science and technology. Of course, scientific thinking appeared and developed independently of capitalism, as is clear, for instance, from the history of mathematics and astronomy. Such outstanding inventions as magnetic compass, fire gun, and book printing date back to pre-capitalistic medieval time. However, scientific discoveries and inventions were rare and casual before the seventeenth century. The turn to systematic scientific and technological research was obviously induced by the onset of capitalism as it pushed forward production for satisfying the arising demand. It was to a great extent a reflection of specific thinking in Protestant cultures: the part of England and Holland in natural sciences and technology is well known. The high prestige of science in those countries was due to the incentive to prosperity which Max Weber interpreted as the generic feature of capitalism and associated with the Genevan thinking.

However, scientists and engineers have no special reasons for liking capitalism being, possibly, among the most hardly exploited groups of population. Researchers in developed capitalistic countries have a few percent of profit from their inventions even if the latter are accepted and implemented. Meanwhile, capitalism would not exist without inventions as it was the developing machinery, or technology advance, that were driving the development of capitalism.

Social significance of rent

As we showed above, rent is a very general phenomenon that appeared long before money and, in a sense, works in animal communities as well. Its common meaning is that an individual benefits from his position which is independent of his efforts. Among people the social position of an individual is justified by tradition or by a legal act, the two being actually equivalent because obeying legal acts is likewise a tradition, more or less formalized depending on the level of civilization.

Reasons why the position of an individual is acknowledged in a community can be different: it can be a chief or a priest in a primitive tribe, a conqueror in a conquered country, an official in a totalitarian state, and, finally, a proprietor. Being an owner is safer than being a chief or a priest who have to validate their position by their deeds from time to time or a conqueror, or an official who can suddenly lose their position. The most stable — and the most abstract — rent is associated with property, which became almost sacred in some cultures. An owner getting rent reaps his regular income only because he possesses some papers, acknowledged by the state, that certify his position. As a rule, nobody cares about the way he obtained those papers or are content with the naive mythology of the classical capitalism. Alexis de Tocqueville admitted in his last years that the community benefits from private property presumably being unable to do without it [Tocqueville, 1856]. Or, more exactly, everybody knows that amateurish attempts to displace rent, without due regard for economic facts, called forth social disasters.

REFERENCES

Marx K. 1865. *Value, Price and Profit*, Speech to the First International Working Men's Association, June 1865, Written: between end of May and June 27, 1865; First published: 1898.

Ricardo D. 1821. *On the Principles of Political Economy and Taxation*, John Murray, London, Third edition (First published in 1817).

Tocqueville A. 1967. *L'ancien régime et la Révolution*, Gallimard, Paris (première édition 1856).

Chapter 12

Unemployment Dynamics

Problem of unemployment

The problem of unemployment is inherent to capitalism as it originates from labor market. The existence of labor market inevitably results in unemployment unless the demand for working hands would exceed the supply for some reasons. This shortage, as well as short supply of any commodity, contradicts the free market constraints (see Chapter 9) and is beyond the scope of this book. Unemployment is evidently related to advance in technology but can exist in the absence of any progress. Technology advance can either increase or decrease the number of jobs. An invention or an innovation makes many jobs useless. Machines supplanted many working hands in the capitalistic England of the late eighteenth and early nineteenth centuries which caused revolts and machine wrecking. Marx made a far-going extrapolation of that tendency of his time and believed ever growing unemployment and pay cut to be the necessary consequences of any technological breakthrough. He viewed technology advance as a self-growing process inherently driven by competition which reduces jobs and allows capitalists to cut down wages ever more. This mechanism called positive feedback, made a basis for Marx's principle of absolute impoverishment of the working class under capitalism and the resulting social shocks (see also Chapter 11). However, Marx's model neglected quality and worked poorly already in his time but was fully disproved in the twentieth century.

Contrary to Marx's understanding of labor productivity increase as the same amount of production but with lesser working hands, technology advance induced quality increase which required even more jobs and greater skills. Another consequence of progress Marx never foresaw was increase in living standards associated with the fact that wage workers are paid for quality: high-quality commodities became ever cheaper and available even for low-paid groups of population. Marx neither could anticipate the appearance of new needs (new (C, Q) planes corresponding to new commodities like the plane in Fig. 5 of Chapter 9) which call forth new jobs and partly cancel the jobs cut caused by innovations.

Of course, measures against unemployment require increase in jobs, and the labor saving approach as a way of decreasing the prime cost of products may induce new job problems. On the other hand, striving for ever growing production of whatever — even high-quality commodities — cause ever growing damage to environment. Below we apply the phase portrait approach to investigate the dynamics of unemployment.

Earned income

The phase plane shows annual earned income in a current year (K) along the K-axis and the income of the same individual in the following year (M) along the M-axis. Assume that we investigate a group of people with uniform business activity which allows us to plot a narrow phase portrait approximated by a curve (cf. Fig. 3 in Chapter 1). Thus, the income in the following year is rather strictly controlled by the income of the previous year, i.e., M is a function of K. One possible phase portrait is plotted in Fig. 1 assuming that people spend a fixed sum D_{min} of their total income D on their living and the remainder on increasing the income. If the income (D_n) of a current year n is above the subsistence limit D_{min}, the value D_n will grow infinitely, ever more every year: $M - D_{min} = q\,(K - D_{min})$, where $q > 1$, wherefrom assuming $K = D_n$, $M = D_{n+1}$ obtain $D_{n+1} = qD_n + (1 - q)D_{min}$ (Fig. 1). This enrichment of a wise investor cannot continue too long for the limitation of resources, and the straight-line segment of the plot gives way to a curve of moderate growth. If the income in a current year is below D_{min}, the income of the following year will reduce to zero, and the individual should die. That was a quite usual end in the time of unlimited capitalism when only philanthropy, if any, could save poor people.

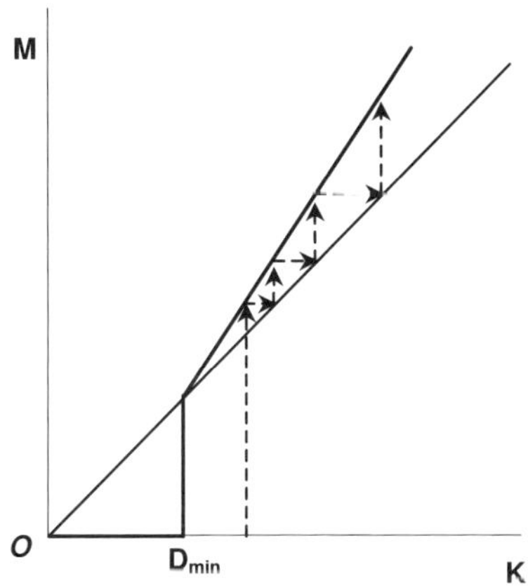

Fig. 1. Possible phase portrait of earned income.

The today's situation rather looks as in Fig. 2, due to efforts of social democrats in developed countries. The infinite enrichment of too successful individuals is restrained by progressive income and estate taxation and support programs for small business which prevent less successive producers from ruin. The right-hand side of the phase portrait (Fig. 2) shows that income growth stops since level 3, and there is thus equilibrium point 3. The left-hand side of the portrait indicates that everybody has some non-zero income: working people have granted floor wages ($D_{\min}$) and unemployed have the right to a fixed dole or welfare (D_w).

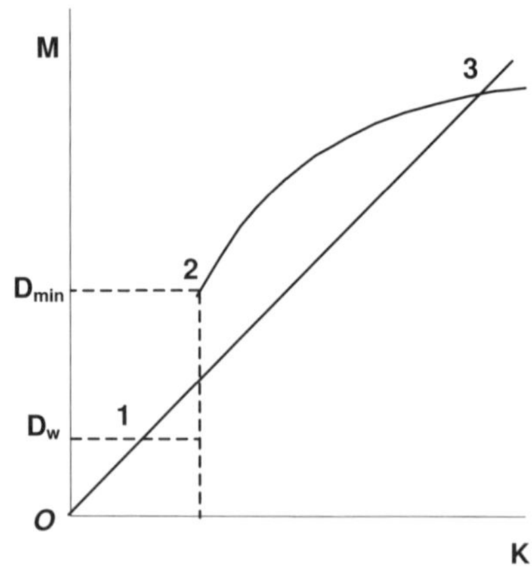

Fig. 2. Phase portrait of earned income changed under the effect of intervention from taxation and reward policy.

Figure 2 illustrates the apparently normal case of floor wages above welfare. In many countries this ratio is carefully maintained, and there is special control that prevents dole recipients from working. In addition to equilibrium point 3 corresponding to an income much above the floor wage, there is a point of unstable equilibrium (point 2) between the domain of growing wages (on the right) and the domain where dole substitutes for wages (on the left), and another equilibrium point (point 1 of stable equilibrium) that marks persistent unemployment. The intricate geometry of this plot is sustained by an army of tax and welfare officers. Unemployment remains irresistible as long as the plot works. The mechanisms that provide this geometry of the phase portrait can be different in different cases, and for different segments of the phase curve. These phase portraits can be plotted on the basis of statistics and are, in a sense, of phenomenological type.

Another, simpler, version of a phase portrait (Fig. 3) implies the absence of unemployment as a persistent phenomenon. The plot would take this shape if everybody is granted a dole slightly above the subsistence limit and there is no income control: anybody is allowed to spend as much as he earns. Progressive taxation is not always necessary as the only stable point of the phase portrait is held back by competition and general limitation of resources. And nobody would starve! The plot in Fig. 3 also implies that unemployed are allowed to work and are not deprived of their welfare if they get a job.

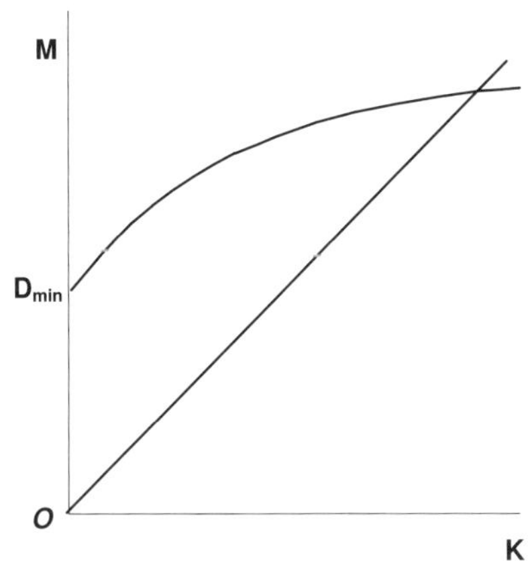

Fig. 3. Phase portrait of earned income in the absence of persistent unemployment.

Of course, the tempting project of Fig. 3 is plausible only in rich economies able to pay doles to anybody. Yet in those countries the project would be opposed as doles has been always treated as charity for poor people and public would hardly welcome the policy of paying welfare to those who are not "poor" as a wise measure against unemployment. On the other hand, the unemployment problem cannot be resolved by a simple change of the phase curve in poor countries where people — whether they have jobs or no — often live below the subsistence.

Phase portrait modeling shows that the problem of total number of jobs coexists with the problem of distribution of the available jobs. A stable class of unemployed may exist or not at any number of available jobs as it arises from the ban for unemployed to work rather than from job shortage. This ban upsets the natural tendency of sharing jobs among all interested people, the tendency that provides labor productivity growth accompanied by decrease in routine and increase in intelligent work and, eventually, makes production less labor-consuming and more environment-friendly.

The real practice can be much more complex than that imaged in simplified phase portraits. Attempts to transform curve 2 into curve 1 by reducing the dole not necessarily drive to the wanted end: People can register themselves as unemployed — to meet the claims of officials and other people who have jobs — but have some odd jobs and in fact live according to the upper part of curve 2, or things can turn in some more sophisticated way.

Note also that people can quit the labor market for business if it can go without starting capital. In the US of the nineteenth century it was easy to become a farmer, which maintained relatively high pays in industry. Russians used to leave for Siberia and the easiest way to quit the labor market in today's Russia is to become a small trader. The possibility for people to get outside the labor market constrains the floor wage D_{min} in Fig. 2 without legal limitations.

As for the shortage of jobs, it can be overwhelmingly cancelled by environmental activities. For instance, road construction which clear the way to new lands giving them more value. A pioneer role in this respect belonged to the projects of Roosevelt's government in the time of New Deal which released the social stress of the 1929–1932 crisis and actually saved the capitalistic economy. But the US leaders haven't learned from that historic lesson. The costly policy of mitigating environmental disasters may arouse complaints of tax-payers. Therefore, these projects require a solid scientific background and much work for public awareness.

Chapter 13

Objects of Nature as Commodities

Environment and property

This chapter deals with a special market, a market of environment objects, as a prerequisite for solving environmental problems we discussed in the first part of our book.

Conservation, or remediation, of nature with its forests, rivers, lakes, and seas and their biota vulnerable to manmade impacts is a matter of both ecology and economy and can be successful only if having reasonable economic underpinnings. Many objects of nature are still taken for granted and believed to be inexhaustible and thus having no price, like air people breathe. Other objects, such as cropland or any agricultural land, have long had their proprietors and sold for a certain price, which in terms of economics means they have been goods. Forests and waters are often no longer free but are appropriated by individuals or companies. In many less civilized regions, however, they are still no man's, or public, or royal property.

Public property is commonly less effectively protected than private property, unless it is guided by religion or taboo as in primitive tribes. Modern man who has lost the primeval wisdom of his ancestors often abuses and damages nature which is the humankind's possession. The unconscious motivation behind the abusive behavior is worth attention by itself. The concept of property historically postdates the primitive tribal attitude that nature is "public belonging". Nobody would doubt that air belongs to everybody and everybody has the right to breathe. An idea to forbid taking water from a river or a source would appear absurd in countries abound in water. For instance, the historic evidence from Ovid says that nobody can hinder the use of water which is for all people[a]. Yet, countries of irrigated cropping have developed a quite different attitude in this respect since old times.

The property for arable land arose as early as agriculture, though communal land tenure held up as late as the modern times in some countries. It existed in medieval Germany, still exists in India, and persisted till late in the nineteenth century in Russia (see also Chapter 9). Russian peasants were against land owning and believed land to be Lord's possession: they accepted landlords to appropriate their labor but not the land. Even nowadays people brought up in similar cultures are often annoyed to see placards announcing private property which bar approaches to the river- or sea-side.

This attitude is by no means strange, as the things nature had created long before humans and without their assistance hardly can be a property of anybody in the sense as a house one has built or a thing one has bought. Appropriation of nature objects by individuals or companies is not the best solution to conservation problems, and this kind of property will hardly exist in the future *"Sane Society"* [*Fromm*, 1955]. Building such a society, however, requires long education efforts to turn the people's thinking from self-interest to moral motives in everything they do, including nature conservation.

Many proprietors of large companies responsible for environment conservation behave like true "communists" taking the objects of nature that have no legal owner for "communal property" and feel easy to damage them. They don't care about the harm they cause to all people, to themselves, and to their children being short-sighted and prodigal in everything which has no immediate pecuniary value. It is reasonable to constrain the right for property by public interests to keep these proprietors from abusing their property. "Wild capitalism" in the sense of absolute right to use and abuse any property no longer exists. In the time of cholera in the 1830s, the London authorities had to impose limitations on the use of private wells. That measure was indispensable to stop the epidemic and was undertaken as soon as the way of infection

---

[a] Quid prohibetis aquis? Usus communis aquarum est. — Ovid, *Metamorphoses*, Book the Sixth. — Lat.

propagation had been understood. Yet, many did not realize that and refused, which is the typical behavior of an owner who repels any interference with his private business.

Nevertheless, elements of nature should have their owners, though with necessary limitations to prevent abuse. Experience shows that environmental measures such as fining or any penalty imposed by national or local authorities are more effective if concern the objects that have an owner interested in their conservation. Economic strategy is the only possible conservation policy, as it will remain driven mostly by property interests in the nearest future. Therefore, it is reasonable to value elements of nature as commodities. Many of them have never had any price before and need at least tentative pecuniary valuation, as it has been done recently in some countries, such as Canada.

Competition for resources

To become commodities in a market, objects of nature should have some quality amenable to numerical evaluation, like land which has had its price since very old times. Let the price of land be denoted as Q and given by

$$Q = DS\frac{100}{p},$$

where DS is the rent (cost gain) of a piece of land, p (%) is the rate of profit in farming in the current practice of a certain country or region. Note that cropland price only partly depends on environment because the owner can make it more fertile by melioration or less fertile by abuse[b]. The fact that people can buy and sell land fits the definition of a free market of nature objects. The quality of timber areas is as easy to value as cropland. Recreation land is a more difficult case. A piece of such land can be valued if this or similar area is already used for recreation as

---

[b] The same approach is applied to manmade property whose value depends on its environment, e.g., houses. We, however, leave aside the cases of manmade objects traded by people for ages as their quality can be defined by their market price and utility according to von Neumann's trading-off procedures (see Chapter 9).

a park in outskirts of a city[c]. Then the rent of its hypothetical owner is estimated by multiplying the mean annual number of tourists by the possible entrance fare people are ready to pay (learnt from a public opinion poll). Such practice already exists in some countries, for instance, in Canada. Finally, forest areas should be valued even if inaccessible for recreation as they contribute to global resources of oxygen, water, ozone, etc.

Imagine an island in which people use forest for recreation. Each area i of this forest, say one acre, has its quality Q_i. Its prime cost is found as total costs S_i for maintenance including costs for preservation (pay to rangers, mitigation of fire hazard and tree diseases, planting, slash removal), roads, communications, etc. Therefore, recreation areas are shown in the phase plane (S, Q) as the points i with the coordinates S_i, Q_i (Fig. 1).

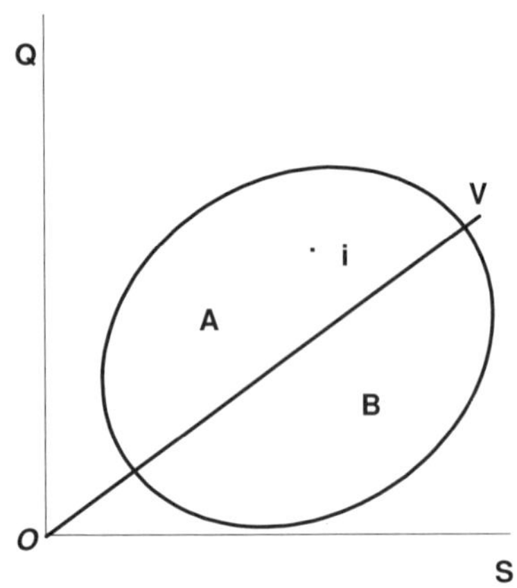

Fig. 1. Phase portrait of recreation land use.

Let the population of the island be N and the recreation norm (the minimum required per capita recreation area) be R. Then the total required recreation area for the whole population is $G = NR$. Assume that the total forest area is larger than G and G is delimited by the straight

[c] For example, the Natural Reserve "Stolby" near Krasnoyarsk in Siberia where picturesque rocks rise among the taiga forest, a very popular place to spend a weekend, climbing rocks or just having a pleasant walk.

line V in Fig. 1 making up the domain A, in which the areas i are selected for recreation, while the areas below V (the domain B) are for other uses. As in the case of cropland, the ratio S/Q constant along V defines the market price of areas in terms of recreation, i.e., assuming that they have no other utility. Forest in the domain B is not used for recreation but the line V goes down if population increases, hence, areas in B make a "standby recreation zone", which preserves forest from cutting. If forest in the island is used as both recreation and timber lands, we arrive at a problem of the economy-environment interaction which can be approached using mathematical modeling.

River valleys offer another illustration to the problem of efficient use of objects with double utility. River valleys[d] can have farming (cropping) or energy generation uses (as water reservoirs for hydro-power plants). Their utility can be measured, as for any commodity, by the monetary cost related to the fixed market price. Let crops from a valley have the utility Q_1 and labor cost S_1, and the utility and labor cost of energy be, respectively, Q_2 and S_2. Then the specific utility (per labor cost unit) of a valley is given by the ratios $\Pi_1 = Q_1/S_1$ and $\Pi_2 = Q_2/S_2$, for farming and energy generation, respectively. River valleys are shown in Fig. 2 as points with the coordinates Π_1, Π_2.

Let all valleys in the east (right-hand side of the sketch) be good for farming and those in the west (left-hand side of the sketch) be good for energy generation. The vertical line is a boundary to the right of which farming land has a high specific utility and the use of all valleys for farming fully satisfies the existing food demand. The horizontal line marks the boundary above which high specific utility belongs to energy generation and the energy demand is fully satisfied if all these valleys are used for power plants.

Thus valleys in the domain A (Fig. 2) are economic for farming only and valleys of the domain B are good only for energy generation[e]. Besides, there is the domain C good for both farming and energy

---

[d] The term *valley* is used symbolically and not necessarily means the all-length valley of a river but rather its part developed (or not) for energy generation uses.

[e] Any consequences of energy generation other than the loss of farming land are neglected for simplicity.

generation where the valleys have equally high utilities in both. Then farmers and energy producers within *C* inevitably are to compete, and the northeastern boundary is to be a troublemaking place. Of course, both parties have to give in and somehow share the domain *C*.

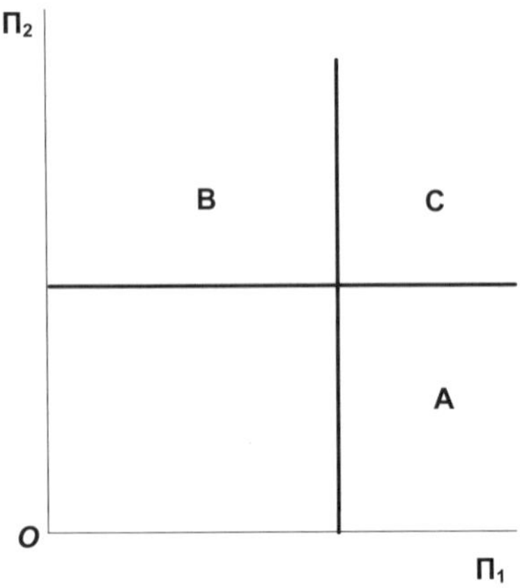

Fig. 2. Sketch "map" of territory with river valleys good for farming or energy generation.

If a part of the domain *C* is given up to energy generation, the demand for crops remains unsatisfied, as it requires *all* valleys to the right of the vertical line (domain *A* plus domain *C*). Then farmers have to develop less economic valleys to the left, and the vertical line shifts leftwards. Similarly, the use of some valleys in *C* for farming can provoke energy shortage and need in using less economic valleys for energy generation; hence, the horizontal line shifts downward. As a result, people arrive at some balance (Fig. 3), hopefully, by means of peaceful bargaining acceptable for both parties. Eventually, farmers occupy the domain *A* and the neighboring domain *K*, while the domains *B* and *L* go to energy producers. Therefore, the domains *K* and *L* are to be divided by a border, and we have to constrain its geometry. Note that there is also the domain *E*, a "wild place" where nobody lives. It will evidently come into use as far as population grows.

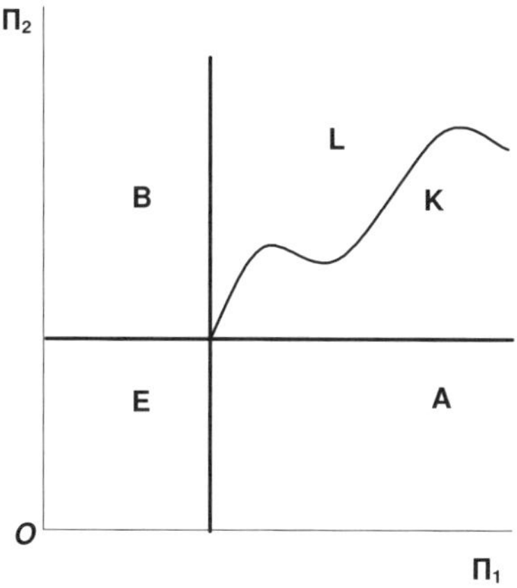

Fig. 3. Sharing valleys between farmers and energy producers.

Sharing land: optimum principle

To set the disputed border between *K* and *L*, the competing farmers and energy producers need a guiding principle acceptable for both. This principle implies that the territory should be shared in a way to fully satisfy the existing local food and energy demands at the least labor costs in both activities.

This formulation would be strange in the context of the common idea of "absolute competition" between producers with quite contrary interests. Yet, it appears natural in terms of the public benefit as a whole. Legislative compliance of private interests to the community requires public consensus on the regulation, including election of those responsible for setting the rules and controls. Coercive and unqualified regulation in economy can lead to costly experiments which pretend to be called "socialism" or, at less imposing intervention of government, to the sway of bureaucracy and suppression of private enterprise.

However, due regard for different interests may yield solutions beneficial for *all* parties concerned and the best possible at the existing environment conditions. This fact, which we illustrate below for the case of valley share can shed new light on the problem of human nature, a popular subject in sociology and philosophy. Many sociologists and philosophers proceed from the idea that "man is malicious" and only pursues his own ends, which rules out fair trade. Of course, unwarranted share of economic and environmental values can drive to wars. The war-oriented "parties concerned" hardly understand that the aggressive behavior is scarcely successful but often turns good for some "third party" or, paradoxically, they are the losers who benefit as they may gain revenge in peaceful economic competition.

We assume that people always bargain peacefully, including the conflict between farmers and energy producers.

First of all, assume that any feasible valley sharing between farming and energy generation provides full coverage of food and energy demands of the local population, corresponding to the amount required by the set market. Indeed, if farming products were in a shorter supply than the marketable output, some people would wish to use valleys less economic for farming and the limits of the domain *A* would shift leftwards (Fig. 3); the limits of *B* are in the same way controlled by energy demand.

Assume, again for simplicity[f], that the utilities and the prices of farming (Q_1) and energy (Q_2) products corresponding to their market price are the same for all valleys. By denoting the farming labor costs for a specific valley as S_1 we introduced its "specific utility" Π_1, i.e, price per unit labor cost:

$$\Pi_1 = \frac{Q_1}{S_1},$$

wherefrom

$$S_1 = \frac{Q_1}{\Pi_1}.$$

Unlike Q_1, the values of S_1 and Π_1 are different in different valleys because environment conditions are different (one can fit the size of areas to equalize their output but cannot change their environment!). Similarly, for energy generation

---

[f] To avoid cumbersome mathematical derivation.

$$S_2 = \frac{Q_2}{\Pi_2},$$

where Q_2 is the same for all valleys and S_2 and Π_2 are different[g].

Consider a simple transaction, trading-off two valleys, which are represented in the figures by points with the coordinates Π_1 and Π_2. Let all farmers working in one valley move to the valley developed by energy producers and the latter move to the farmer's valley. As, according to our assumption, all valleys provide equal food or energy supply, this transaction does not contradict the condition of full coverage of the demand. We keep the same designations for the first valley (S_1 and Π_1) but introduce S_1' and Π_1' for the valley now under farming. Then, the labor cost increment (negative or positive) as a result of the exchange is $\Delta S = S_1' - S_1$. We had $S_1 = Q_1/\Pi_1$ for the first valley and $S_1' = Q_1/\Pi_1'$ for the other valley (Q is the same in all valleys), then

$$\Delta S_1 = S_1' - S_1 = \frac{Q_1}{\Pi_1'} - \frac{Q_1}{\Pi_1} = Q_1\left(\frac{1}{\Pi_1'} - \frac{1}{\Pi_1}\right).$$

Similarly, the labor cost increment for energy producers who move to the farmers' valley is

$$\Delta S_2 = S_2 - S_2' = \frac{Q_2}{\Pi_2} - \frac{Q_2}{\Pi_2'} = Q_2\left(\frac{1}{\Pi_2} - \frac{1}{\Pi_2'}\right).$$

The trade-off is evidently feasible only if both competing parties benefit (note again that we mean only voluntary transactions). It is possible in two cases: (i) if the exchange reduces labor costs of both parties, i.e., if ΔS_1 and ΔS_2 are both negative, and (ii) if one side gains, i.e., one increment of labor cost is negative and the other is positive. Let, for instance, farmers gain and energy producers lose immediately in the exchange, i.e., $\Delta S_1 < 0$ but $\Delta S_2 > 0$. One would think that energy people

---

[g] Note that the share sizes selected to provide equal Q_1 do not necessarily mean equality in Q_2, and the above assumption thus concerns a particluar case of the problem. The general case is solvable — with the same approach — but we confine ourselves to the particular case for the sake of derivation simplicity.

would never accept such trade-off. However, consider a particular case when

$$\Delta S_1 + \Delta S_2 < 0$$

(Note that this inequality fulfills in the former case as well, when both summands are negative). Then the absolute ΔS_1 exceeds ΔS_2 (Fig. 4):

$$|\Delta S_1| > \Delta S_2 .$$

It means that having a gain greater than the loss of energy producers, farmers can spend a part of this gain to compensate the loss to energy people, even with surplus, and thus both benefit. This is an example of a fair transaction. The conflict is resolved in the same way if energy people gain and farmers lose. Both cases are summed by the same inequality $\Delta S_1 + \Delta S_2 < 0$.

Fig. 4. Exchange of valleys between farmers and energy producers.

If this condition fulfills, both parties benefit — if the reward is appropriate — and the trade-off is practicable. However, the exchange means that the total labor costs (for farming and energy generation) are reduced, as the decrease in farmers' labor cost is greater than the labor cost increase of energy producers. Thus, if S_{I} is the total labor cost for farming and S_{II} is the total labor cost for energy generation, at $\Delta S_1 + \Delta S_2 < 0$, the trade-off reduces the sum $S_{\mathrm{I}} + S_{\mathrm{II}}$.

Now remember that only farming and only energy generation are economic in the domains A and B, respectively (Fig. 3), and the problem is to share the disputed area where both activities are equally economic, i.e., to set the border between K (farming area) and L (energy generation area) (Fig. 3). One can expect that farmers and energy people eventually arrive at trading-off as described above. It is mutually beneficial and possible if $\Delta S_1 + \Delta S_2 < 0$.

Obviously, the greater the negative left-hand side of the inequality the more profitable the trade-off, as both parties benefit. Trading stops when the gain reduces to zero, and the border between the farming and energy generation areas becomes finally set. The last transactions evidently occur in the immediate vicinity of the border where $\Delta S_1 + \Delta S_2 = 0$.

Figure 5a demonstrates the case when it is reasonable to trade-off the farming area a along the border on the side K for the energy-generation area a' along the border on the side L.

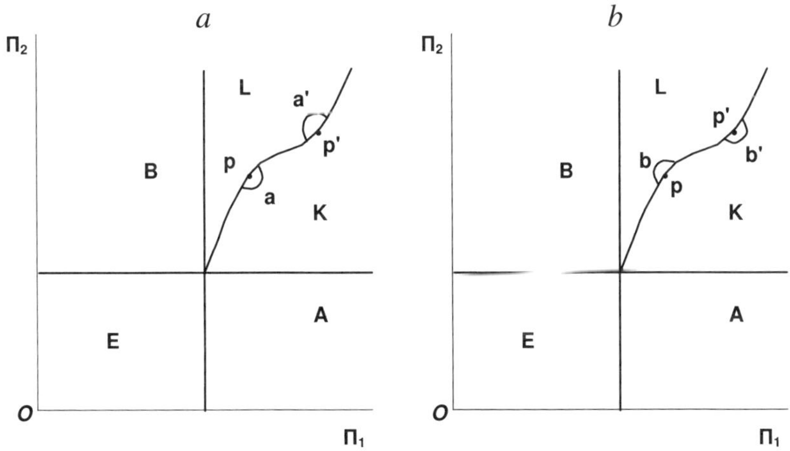

Fig. 5. Optimal dividing the territory into farming and energy generation zones by exchange of farming and energy-generation areas along the border, for inequality (α) in (*a*) and for inequality (β) in (*b*).

Substituting the above equations for ΔS_1 and ΔS_2 into $\Delta S_1 + \Delta S_2 < 0$ gives

$$\left(\frac{Q_1}{\Pi_1'} - \frac{Q_1}{\Pi_1}\right) + \left(\frac{Q_2}{\Pi_2} - \frac{Q_2}{\Pi_2'}\right) < 0. \qquad (\alpha)$$

This is the practicability condition for the transaction, and we now search for valleys along the border between K and L which satisfy this condition. Inequality (α) relates the coordinates of the points p (Π_1, Π_2)

and p' (Π_1', Π_2'). It is reasonable to search the points p and p' fitting (α) along the very border curve (we noted that the last transactions occur near the border). If such points do exist (Fig. 5*a*), the exchange of areas in Fig. 5*a* can reduce the total labor costs $S_I + S_{II}$. We select the sizes of areas a and a' small enough to bring them closer to the points p and p', respectively. Note that the valleys are small relative to the areas and the areas are small relative to the whole "map" in Fig. 5*a* that covers a territory with a great number of valleys. With the valley sizes in the areas a and a' approaching the points p and p', inequality (α) is still valid (the left-hand term is for a and the right-hand term is for a'), and then any valley within a is exchangeable for any valley within a'![h] It thus remains to select the sizes of the areas a and a' near the points p and p' such that they include equal numbers of valleys and trade-off all valleys in a for all valleys in a'. Then the total cost $S_I + S_{II}$ will reduce as noted above.

As a result, the K/L border retreats near p (viewed from the side K) giving up the area a of the domain K and advances near p' taking up the area a' of the domain L. Therefore, it is possible to reduce the total labor costs if some points on the border curve have the coordinates that fit inequality (α), and the market still covers the demands in both food and energy, as this condition was fulfilled in the transactions we described.

However, the sum $S_I + S_{II}$ can reduce also if some pair of points p and p' on the border curve satisfies the contrary inequality

$$\left(\frac{Q_1}{\Pi_1'} - \frac{Q_1}{\Pi_1}\right) + \left(\frac{Q_2}{\Pi_2} - \frac{Q_2}{\Pi_2'}\right) > 0. \qquad (\beta)$$

Figure 5*b* shows the energy generation area b along the border on the side L and the farming area b' along the border on the side K. Trading-off the two areas drives at $S_1 = Q_1/\Pi_1'$, $S_2 = Q_2/\Pi_2'$ for b' and $S_1 = Q_1/\Pi_1$, $S_2 = Q_2/\Pi_2$ for b. Then ΔS_1 is

$$\Delta S_1 = \frac{Q_1}{\Pi_1} - \frac{Q_1}{\Pi_1'},$$

and ΔS_2 is

[h] It is obvious from the above how this exchange is provided by the inequality $\Delta S_1 + \Delta S_2 < 0$ tantamount to (α).

$$\Delta S_2 = \frac{Q_2}{\Pi_2'} - \frac{Q_2}{\Pi_2},$$

Two latter values differ from those in (β) only in the sign of the respective terms, hence for the exchange of b and b',

$$\Delta S_1 + \Delta S_2 < 0.$$

If there is a pair of points on the border that fit inequality (β), the total costs $S_I + S_{II}$ can again reduce by moving the border as in Fig. 5*b*! (Sec where the border retreats and advances in Fig. 5*b*). What does this mean? If neither (α) nor (β) are possible for any pair of points along the border, the sum $\Delta S_1 + \Delta S_2$ is zero at these points, or for any pair of points p and p' on the border

$$Q_1\left(\frac{1}{\Pi_1'} - \frac{1}{\Pi_1}\right) + Q_2\left(\frac{1}{\Pi_2} - \frac{1}{\Pi_2'}\right) = 0. \qquad (\gamma)$$

This equation allows us to constrain the geometry of the border between zones of competing land uses: it turns to be a hyperbola (Fig.6)[i].

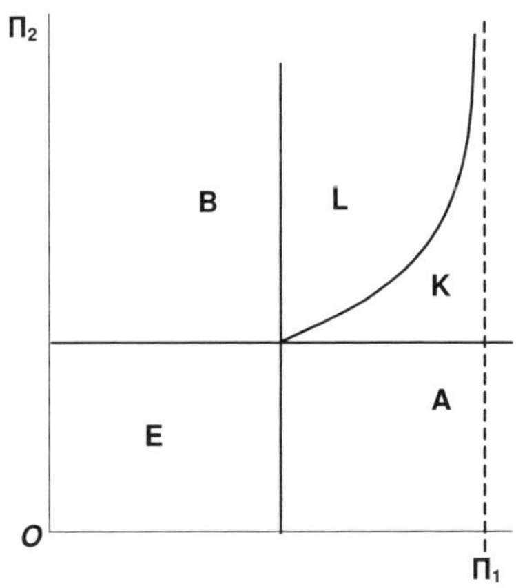

Fig. 6. Finally achieved optimum geometry of the border between zones of competing land uses.

[i] See the proof below in the end of the chapter.

This means that no trading-off is possible any more along the border. The border becomes fixed when all profitable transactions are over. Equation (γ) does not let further decrease in the sum $S_{\rm I} + S_{\rm II}$. Therefore, we actually have arrived at sharing which brings this sum to the minimum[j].

Thus, the spontaneous activity of free market eventually brings to the same end as solving the optimization problem formulated for the sake of public good. This illustrates the idea that free market actually provides the same solutions as the authentic optimum planning (see Chapter 9). Adam Smith, the founder of the economic science, understood this long before the advent of mathematical modeling in economics. He intuitively came to the discovery we have just illustrated, and his contemporaries called it 'Adam Smith's optimism'. He discovered that people benefit the community around them simply by acting solely in their own self-interest, without conscious regard to community service[k], which he called the "invisible hand" of free market[l].

There is no paradox in it: The trading activity induces competition which raises the energy of self-interest. Its noxious imprint on the personality of tradesmen and on the community as a whole is a different matter. Adam Smith, an economist and a philosopher, understood this much better than his successor advocates of free market. He only told that free market was able to provide the best labor productivity and, hence, national wealth but he was by no means optimistic about the way of sharing and using that wealth.

The question naturally arises why don't people change free market (which is by the way no more free nowadays) for direct optimum planning? True large-scale optimum planning is a challenge to foresee human thinking and behavior which is far beyond the understating of

---

[j] The approach is related to calculus of variations. We applied a number of simplifying assumptions to avoid procedures requiring higher mathematics necessary for more realistic cases of varying the border curve of Fig. 5.

[k] Smith wrote: "It is not from the benevolence of the butcher, the brewer, or the baker, that we can expect our dinner, but from their regard to their own interest".

[l] ... "he intends only his own gain, and he is in this, as in many other cases, led by an invisible hand to promote an end which was no part of his intention..." [Smith, edition 1994].

those who has ever smattered such planning. Therefore, it is market that will remain responsible for optimization in national economy in the nearest future. This does not mean, however, that mathematical optimization solutions are useless. On the contrary, they respond to many essential particular problems urgent for survival in the high-technology world.

The fact that the border curve between K and L is a hyperbola (Fig. 6) can be proven with straightforward tools of analytical geometry.

Equation (γ) includes the coordinates of two points on the sought curve $p(\Pi_1, \Pi_2)$ and $p'(\Pi_1', \Pi_2')$ (Q_1 and Q_2 are constants that define the valleys' output); equation (γ) fulfills at any choice of these points. Let the point p' (Π_1', Π_2') be fixed and the point p run along the border curve. Then the coordinates Π_1, Π_2 fit equation (γ) where all other variables are constants. Rewriting this equation to

$$\frac{Q_2}{\Pi_2} - \frac{Q_1}{\Pi_1} = \frac{Q_2}{\Pi_2'} - \frac{Q_1}{\Pi_1'}$$

and denoting the right-hand side as a, Π_1 as x, and Π_2 as y gives

$$\frac{Q_2}{y} - \frac{Q_1}{x} = a,$$

or

$$Q_2 x - Q_1 y = axy.$$

To simplify this equation, move the x and y axes for the distances x_0 and y_0:

$$x = x_0 + x', y = y_0 + y',$$

where x' and y' are the new coordinates of the point p. Thus

$$Q_2 x' - Q_1 y' + Q_2 x_0 - Q_1 y_0 = a(x' + x_0)(y' + y_0),$$
$$ax'y' + x'(ay_0 - Q_2) + y'(ax_0 - Q_1) = Q_2 x_0 - Q_1 y_0 - ax_0 y_0.$$

We select the shifts x_0 and y_0 to zero the parentheses on the left, substitute these numbers into the right-hand side, and denote the obtained number as ac. Reducing by a gives the hyperbola equation:

$$x'y' = c \text{ (or } y' = c/x').$$

This is the sought curve that divides the right-hand top corner into the domains *K* and *L*. The hyperbola cannot cross the borders of the domains *A* and *B* as it divides valleys of different utilities and the domains *A* and *B* include only valleys of the same utility (farming in *B* and energy generation in *A*). Hence, it passes through the vertex *C* of the rectangle.

REFERENCES

Fromm E. 1955. *The sane society*, Rinehart & Company, New York, 370 pp.
Smith, A. 1994. *An inquiry into the nature and causes of the wealth of nations*, Random House Inc., New York–Toronto.

Chapter 14

Long-term Motivation

In this chapter we investigate the long-term motivation of proprietors measured by their concern about the next generations. Proprietors in our meaning include a broad range from an individual to a state. The objects of property can be manmade things, like buildings – houses or palaces, – bridges, factories, etc., or environmental wealth, such as land or forest. Land and forest have been bought and sold for market prices since long ago in the civilized countries. As we wrote in Chapter 13, many things people receive from nature free of charge can become market commodities. Moreover, consistent pecuniary valuation of environment objects and assigning them to proprietors concerned about their preservation, like private persons or companies are concerned to preserve any their belongings, appears to be the only way to save nature given the existing social attitudes. The state can be such a proprietor if it proves effective in environment protection. In Canada many environmental values are the property of the queen, or actually of the state, and this is a good solution because the Canadian administration takes care of nature without trespassing on people's rights. On the contrary, nature owning by the state should not be allowed in countries where the "national property" in fact has no protection but is abused by officials.

Unlike many manmade products, especially, the necessaries, objects of environment such as land or forest are used for a period far exceeding a human life. A loaf of bread, a tube of toothpaste, or a pair of shoes are of very short run. A car can serve for years though its owners prefer to have new fashionable models every couple of years. Buildings or factories most often remain in use for a time comparable to or longer than people's life. The lifetime of a house is most often longer than its owner's life; a factory enterprise can be operated by many generations but the equipment requires upgrading at least every decade.

In view of proprietors, whose behavior has been driven by aspiration for cutting up a profitable property since far pre-capitalistic times, houses and especially land are more solid sources of profit than factories. A factory is a very troublesome and not very safe property, which one is never sure to leave well running to his heirs because many industries are coming out of date very soon nowadays.

Parameter of egoism

The profit rate or mean annual percentage of revenue relative to laid-down capital, has always been lower in farming than in industry. A today's factory owner has, in the average, a cost gain of 10–11% of the factory worth, whereas a land owner receives 6–7 % of the land value. The causes of this difference challenged many economists. Marx was vainly searching for explanations over tens of pages in the third book of his *Capital* (which he never ended).

We put aside the economic reasons and are trying to discuss the phenomenon as a matter of people's thinking and motivation. Economists, by the way, have always invoked the psychological context of people's behavior in trade affairs. We did the same in Chapter 9 and the motives we discuss can be easily found empirically and are independent of psychological theories.

Let the mean profit rate in a society be p' % in industry and p'' % in farming. Empirically, $p' > p''$, or $p'/p'' > 1$, or

$$p'/p'' = 1 + \varepsilon,$$

where ε is positive. Assume that a greater ε corresponds to greater concern of individuals in a given generation for their descendents. To put it different, those who buy land instead of factories are content with a lower rent for the sake of well-being of their children, assuming they treat land as a safer property than factories and want to cut up well. Then, ε measures proprietor's concern for his descendents: greater at greater ε and less at lower ε. At $\varepsilon = 0$, i.e., at $p' = p''$, a proprietor does not care whether to own land or a factory, and the higher profit from land is assumed to be provided by its abuse. Yet such absolute egoism has never existed in history, and even today $p' > p''$. Estimating ε can be difficult for countries of former or present command economies where land and factories were or remain nationalized and are out of market, though some are known to have very low concern for the future generations.

A proprietor[a] is always subjective in evaluating the quality of his property and the assigned value may differ from the current market price.

Being aware how difficult and vulnerable any numerical estimates of private experience are, we have not approached them before in our book though placed the behavior of market agents in the context of people's mind. Now we assume that a proprietor assigns his property a time-dependent quality $Q(t)$ and this dependence is plotted in Fig. 1.

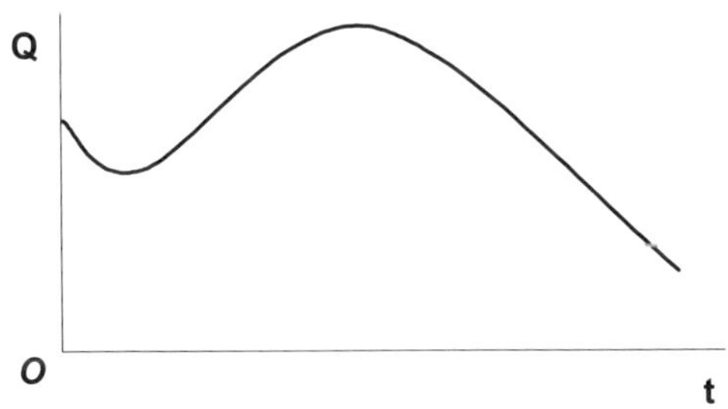

Fig. 1. Time dependence of the quality of property viewed by its proprietor.

The quality of a durable commodity normally decreases for the life time of an individual, because of wearing or becoming outdated ($Q(t)$ is zero in objects that have passed out of use). Or, a proprietor can increase the quality of his property, for instance, make his land more economic by planting forest.

---

[a] We use the term "proprietor" as the singular making no distinction between individual and cooperative proprietors as any property is eventually handled by people, whether an individual or a company, more so that the principal motivation is more or less the same for all individuals in a company.

We operate with qualities averaged over a period of time. Figure 2 shows a $Q(t)$ plot between some initial time t_i and final t_f. The area S below the plot constrained by the t axis from below and by the intervals $t = t_i$ and $t = t_f$ on the sides equals the area of some rectangle (cut by a dotted line on its top in Fig. 2) with the same base and the height taken for the average Q_{av} of the function $Q(t)$ over the interval (t_i, t_f)[b].

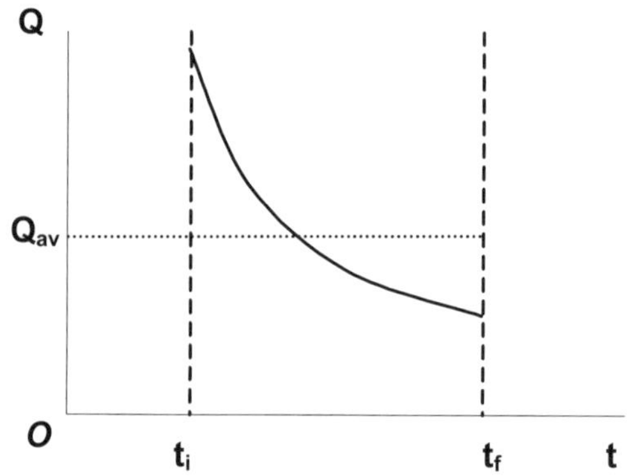

Fig. 2. Time dependence of quality taken within an interval between some initial time t_i and final t_f.

Hereafter we consider objects of nature whose lifetime exceeds the human life span T, such as cropland and timber forest. Forest appears more interesting in terms of long-run motivation of proprietor's behavior, as the yield of cropland annually depends on the effort of its owner whereas forest grows too slowly to be profitable in a short run. A pine grows for 100–150 years to become grand timber, and the owner of a pine forest cannot expect to gain profit himself, the income will go to his grand-children. Nevertheless, planting forest is a usual activity in Europe and in the US. According to some — possibly too optimistic — estimates, the US currently has more forest than hundred years ago and

---

[b] The reader familiar with elements of integral calculus can easily check that Q_{av} is the limit of the arithmetic means $(1/n)\left(Q(t_1)+Q(t_2)+\ldots+Q(t_n)\right)$, where t_1, t_2, … t_n are close successive times from t_i to t_f.

almost all forest in Europe, except for that in the Carpathian mountains, originates from manmade plantations.

The immediate motivation behind planting forest is to sustain and increase the market price of timber which the owner can always sell to cover his costs. However, why a product that cannot be used today would keep its worth?

Let a proprietor (of timber forest in this case) assign his property some quality at any time t assuming that this quality will persist after his death. Therefore, the function $Q(t)$ never reaches zero even after the end of the period T though it is normally a decreasing function (Fig. 1). The function can even increase if the owner takes the proper care of his forest or plants more trees.

Let Q_0 be the average quality of timber over the life of its owner (from $i = 0$ to $t = T$), Q_1 its average quality over the life of the owner's son (from $t = T$ to $t = 2T$), and Q_2 that over the life of the owner's grandson, etc., provided they inherit the forest successively. Mind that Q_0, Q_1, and Q_2 are the estimates of the future timber quality assigned by the current owner, which are evidently personal. However, they can become objective if all timber owners in a country set approximately equal values at a given time. Assume that Q_0, Q_1, and Q_2 make up a descending sequence, i.e., the owner of timber values its quality for the life of his son lower than for his own life, the quality for the life of his grandson lower than for the life of his son, etc. The simplest inference is that these numbers make up a descending geometric series: Q_1 is k times lower than Q_0, Q_2 is k times lower than Q_1, etc., where $k > 1$.

Obviously,

$$Q_1 = \frac{1}{k} Q_0, \ Q_2 = \frac{1}{k} Q_1 = \frac{1}{k^2} Q_0, \ldots;$$

finally, $Q_n = \frac{1}{k^n} Q_0$ for any integer positive n.

Now we express this relationship via the time t rather than in terms of generations. If an average human life spans the period T, n generations span the time nT. If $t = nT$, $n = t/T$, and Q in t years, or Q_n, is $Q(t)$. Then,

$$Q(t) = \frac{1}{k^n} Q_0 = k^{-\frac{t}{T}} Q_0.$$

$Q(t)$ is an exponential time function with the base k and the exponent $(-t/T)$. To this point we have assumed that the time t is divisible by T, i.e., consists of an integer number of generations, and for each generation we considered the average quality of timber over the life periods of the owner, his son, grandson, etc.

Now, to generalize, we proceed from the average timber quality for successive generations assigned by the current owner to quality at any time t assigned by an average proprietor. Assume that $Q(t)$ is given by the same exponential function we used above. This assumption obviously requires empirical testing, namely, explaining the way to define k. Therefore, if the average timber quality assigned by the owners at the time $t = 0$ is Q_0, the average quality of the same timber at the time $t > 0$ is

$$Q(t) = Q_0\, k^{-t/T}.$$

The transition from time average to the average over a group of people may appear somewhat arbitrary, but this is a normal procedure in statistical physics[c].

Those who use exponential functions prefer to deal with a fixed base (rather than the arbitrary base k in the previous equation). Thus we can easily pass to some base a. Indeed, to be valid, the equation (time identical)

$$k^{-t/T} = a^{-\lambda t}$$

only requires that the logarithms of both sides were equal in the base a:

$$-\frac{t}{T}\log_a k = -\lambda t\,.$$

Then

$$\lambda = \frac{\log_a k}{T}\,.$$

Thus $Q(t)$ can be written as $Q_0 a^{-\lambda t}$ with any positive base a and λ defined as above.

---

[c] We mean the so-called ergodic hypothesis. The reader strange to physics can safely miss this note as our hypothesis can be checked by practice.

Hereafter we use exponential functions with the standard base $e = 2.71828\ldots$, so-called Napier's number[d], which is common in natural sciences. Assume that $a = e$ in the equations above and denote $\log_e$ as ln (natural logarithm). Then

$$Q(t) = Q_0 e^{-\lambda t},$$

where $\lambda = \dfrac{\ln k}{T}$.

At $t = 0, T, 2T, \ldots$, evidently $Q(t) = Q_0, Q_1$, and $Q_2, \ldots$ (as $e^{\ln k} = k$). Thus we arrived at the same values of quality in one, two, and more generations, whereby the ratios Q_1/Q_0, Q_2/Q_1, … = $1/k$. The factor k increases proportionally to λ as $e > 1$ and, hence, $\ln\lambda$ is an increasing function. Therefore, the higher λ corresponds to less concern for descendents, and their interests are absolutely neglected at infinitely large λ. So λ can be called the "egoism parameter".

Naturally, the quality the owner assigns to his property is proportional to the expected gain. Indeed, living standards are commonly assessed against a list of consumed products and the range of commodities one can buy is determined by his income. We confine ourselves to this definition of quality and assume that the assigned quality of property is measured by the rent it can yield for the period of the proprietor's life. On the other hand, the social estimate of quality corresponds to the maximum derivable profit. This profit obviously cannot be lower than the market price of the property which otherwise nobody will buy; neither it can be higher than the price as sale in this case would be a loss for the proprietor. Then, the social evaluation of the quality of some property corresponds to its market price.

The two estimates may differ as the proprietor can value his property high for some more reasons than a derivable income, for instance, willing to leave it well preserved for his heirs. Consider again the example of a private timber forest. The owner of an area under timber assigning it the average quality Q_0 cannot allow himself to spend the whole market value of his timber on his own needs and allots

[d] John Napier (1550–1617) invented logarithms. The base e is preferred because exponential functions with this base do not change during differentiation.

$Q_1 + Q_2 + \ldots$ to his heirs. The quality he leaves for himself is proportional to his annual profit, or his rent p''. On the other hand, the owner of a factory of the same market value $Q_0 + Q_1 + Q_2 + \ldots$ may never care about his descendents being aware how fast the equipment becomes outdated, and spend all the profit from his factory on himself. Then his rent p' will be proportional to the sum above, and the ratio p'/p'' is

$$\frac{p'}{p''} = 1 + \varepsilon = \frac{Q_0 + Q_1 + Q_2 + \ldots}{Q_0} = 1 + \frac{Q_1}{Q_0} + \frac{Q_2}{Q_0} + \ldots$$

Therefore,

$$\varepsilon = \frac{Q_1}{Q_0} + \frac{Q_2}{Q_0} + \ldots$$

or, with the equation for $Q(t)$:

$$\varepsilon = e^{-\lambda t} + e^{-2\lambda t} + e^{-3\lambda t} + \ldots$$

The right-hand side is a geometric series with the ratio $e^{-\lambda T}$. Since $\lambda > 0$, $e^{-\lambda T} < 1$, and the sum is found as

$$\varepsilon = \frac{e^{-\lambda T}}{1 - e^{-\lambda T}}.$$

Then the egoism parameter λ can be expressed via ε as

$$\left(1 - e^{-\lambda T}\right)\varepsilon = e^{-\lambda T}, \qquad \varepsilon = e^{-\lambda T}(1 + \varepsilon),$$

$$e^{-\lambda T} = \frac{1 + \varepsilon}{\varepsilon},$$

hence,

$$\lambda = \frac{1}{T} \ln \frac{1 + \varepsilon}{\varepsilon}.$$

Thus the egoism parameter is controlled by the profit rate (or rent) yielded by industrial and land property existing at a given time in a given place[e]. This parameter allows us to assess the attitude to the future generations in a culture at any stage of its history, as the profit rate that existed in the past can be most often learned from statistics or archives.

The extreme cases are easily found in history. European travelers who met people of primitive tribes discovered the quality estimates free from distinction between the property of the current and future generations, as all belongings of a tribe were its property while a tribe was meant as a totality of all past, present, and future generations. In this case $Q_0 = Q_1 = Q_2 = \ldots$, the egoism parameter λ is zero, and the total quality is infinite. Of course, nobody can buy forest from such a tribe, their forest can only be seized by killing all people! Apparently puzzling changes in the attitude of natives toward the hosted Europeans always arose when the latter misunderstood the values inherent in the strange cultures. Konrad Lorentz wrote about first German explorers of New Guinea who were friendly welcomed by the natives but then incurred their great anger by cutting some large tree which turned to be a tribe's sacred thing.

The present Western society represents the other extreme relative to the primitive cultures, as it cares little about the coming generations. The λ parameter in this society is very high, Q_1 is much lower than Q_0, and the following Q_n are vanishing. We attribute this tendency to the ongoing global degradation of the Western culture[f] dominated by a hedonistic attitude of people unable to sacrifice their momentary pleasures. Yet, the common land/industry rent ratio persists, and land, forest, and other environmental values are still considered safer for investment.

A study of time-dependent λ changes in different countries would be a stimulating contribution to the historic account of our culture.

The exponential function that measures the "decreasing interest" of proprietors in their property appears to be at odds with the initial conditions it was derived from. In fact each individual at every time of his life values the quality of his property on the basis of trade and applies

---

[e] Note that the forest and land rents were assumed equal. A more thorough consideration would require taking into account their possible difference. Yet, the same derivation can be applied to the land property, perhaps, with a slightly different rent.

[f] We use the concept of Western culture meaning all cultures that originally stem from the Ancient world and Christianity (all Europe, United States, and Russia).

the correction $e^{-\lambda T}$ only in the case of his descendents. However, the exponential function actually works somehow also during the life of the individual, as he values the current full-grown timber forest higher than that going to grow in forty years. Mind yet that we passed from a time average to a group average in the function definition. The average value accepted in a group of timber owners makes the market price of different tree species, and any proprietor discovers the realistic existence of λ as soon as he intends to sell his timber. The trade will inherently depend on the age of trees and planting efficiency.

Thus the generalization of passing from the geometric series Q_0, Q_1, Q_2, … to an exponential function works in the conditions not meant originally (t originally was assumed divisible by T). This is however typical of all appropriate generalizations.

The law of exponential decrease is common to physics, for instance, in the case of radioactive decay. The mass $M(t)$ of a decayed radioactive element remaining at the time t is given by $M_0\ e^{-\lambda T}$, where M_0 is the initial mass and λ differs in different elements. The time t_m in which a half of M_0 decays is called element's half-life and is found from

$$M(t_m) = M_0/2$$

In the same way, the function $Q(t) = Q_0 e^{-\lambda T}$ describes the evolution of property value in owner's view and the time in which this quality becomes twice as low is given by

$$Q(t_m) = \frac{1}{2} Q_0$$

Therefrom

$$Q_0 e^{-\lambda t_m} = \frac{1}{2} Q_0, \qquad e^{-\lambda t_m} = \frac{1}{2},$$

and, finding the logarithm with respect to the base e gives

$$\lambda t_m = \ln 2, \qquad t_m = \frac{\ln 2}{\lambda}.$$

ln 2 being about unity ($e = 2.71\ldots$), $\lambda \approx 1/T$. Thus, $\lambda \approx 0.01$ for people who value the interests of their children to a half of their own interests. This number appears underestimated for our times.

After a year from the initial time t (taken in years), the quality $Q(t)$ becomes $Q(t+1)$, and the quality decrease is

$$\frac{Q(t+1)}{Q(t)} = \frac{Q_0 e^{-\lambda(t+1)}}{Q_0 e^{-\lambda t}} = e^{-\lambda}.$$

Thus λ is the negative natural logarithm of the annual quality decrement.

Planting forest

It appears curious to investigate the parameter λ, originally derived from profit rate, in terms of timber trade and to check the practicability of planting forest against this parameter.

Let τ be the time of natural forest renewal, i.e., mature forest grows in place of cut forest in τ years. For many trees τ is of the order of one hundred years: 100–150 years for pine, 40–50 years for birch and some other fast-growing trees; such trees as oak grow for hundreds of years. A glade left as is after cutting pines becomes first covered with birch, whereas pine starts growing forty or fifty years later. Birch forest is even good for young pines as it protects them from heat and dewatering, but the species change takes a long time, and full-grown pines appear in 200–250 years instead of 100–150 years. Acceleration of forest renewal is obviously profitable for the owner. He values the quality of mature timber expected to grow naturally in τ years as

$$Q(\tau) = Q_0\, e^{-\lambda T},$$

where Q_0 is the current quality of full-grown timber over one acre. Mind that these quality values are objective as they correspond to the current trade practice, and can be expressed directly, say, in dollars.

Assume that the time τ can reduce to a time $\Delta\tau$ years shorter due to manmade planting and special care. Then timber becomes full-grown in (τ – Δτ) years. An additional advantage of manmade planting excluding the birch "middlemen" is that it yields uniform grand timber, valued as

$$Q(\tau - \Delta\tau) = Q_0 e^{-\lambda(\tau - \Delta\tau)}.$$

The quality (value) gain relative to naturally grown forest is

$$\Delta Q = Q_0\left(e^{-\lambda(\tau-\Delta\tau)} - e^{-\lambda\tau}\right) = Q_0 e^{-\lambda\tau}\left(e^{\lambda\Delta\tau} - 1\right)$$

(where the minuend is greater than the subtrahend).

In order to estimate how much the timber owner gains from planting, note again that τ for high-grade tree species is hundreds of years whereas growth acceleration Δτ is at most thirty to forty years. Assume that τ and Δτ are invariable, i.e., consider the same tree species and the same care and planting technologies. Then the previous equation for ΔQ is a function of the single variable λ which we denote as η(λ) (Fig. 3). Obviously, η(0) = 0, and both $e^{-\lambda t}$ and $e^{-\lambda(\tau - \Delta\tau)}$ tend to zero at increasing λ, and η(∞) = 0.

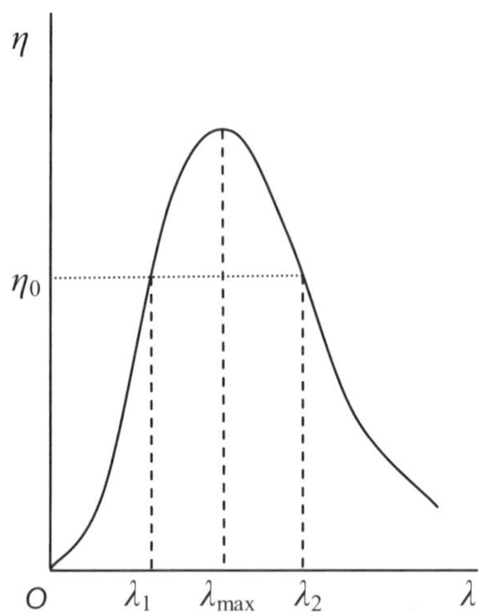

Fig. 3. Profit from forest plantation as a function of egoism parameter.

It can be shown that the greatest η, or the greatest gain is at

$$\lambda = \lambda_{\max} = \frac{1}{\Delta\tau}\ln\frac{\tau}{\tau - \Delta\tau},$$

and the highest η is

$$\eta_{\max} = Q_0\left(1 - \frac{\Delta\tau}{\tau}\right)^{\frac{\tau}{\Delta\tau}}\frac{\Delta\tau}{\tau - \Delta\tau}.$$

Two latter equations are easily obtained by differentiation. The function $\eta(\lambda)$ becomes zero at $\lambda = 0$ and at $\lambda = \infty$ and reaches its maximum at the point $\eta'(\lambda) = 0$, wherefrom we obtained the above $\lambda_{\max}$. Substituting it into the equation for $\eta(\lambda)$ gives the equation for $\eta_{\max}$.

Therefore, the function $\eta(\lambda)$ increases from 0 to $\eta_{\max}$ and then decreases to zero (at λ tending to infinity).

The effectiveness of planting is controlled by its costs. If planting and care cost η_0 dollars per acre, planting pays off only at $\eta(\lambda) > \eta_0$ (Fig. 3), i.e., if the value gain for mature timber exceeds the costs for planting and care. Thus planting is economic only within the domain (λ_1, λ_2) in which $\eta(\lambda) > \eta_0$ (Fig. 3).

People do not plant forest at too low λ (below λ_1), which corresponds to the practice of a primitive tribe, as planting can add nothing to the value of forest meant invariable for the current and future generations. Planting is neither practiced at too high λ (above λ_2), as people in a community with a high egoism parameter do not care what grows after them. Planting fortunately continues in Western countries (but has almost stopped in many countries, for instance, in Russia). The above equations can be applied to investigate the problem of planting at known λ. Such studies are underway at the Institute of Forest of the Russian Science Academy in Krasnoyarsk.

However, the value of forest cannot be measured merely by timber trade. Forest provides people with oxygen and water, not only its owners but all people. This environmental worth is not accounted for by the above theory implying that owners value their forest only in terms of their own profit. The property right in the case of forest (and other elements of environment!) is actually limited by the very fact that no proprietor can appropriate and destroy (say, by cutting his forest) the goods of nature shared by all people who are their natural proprietors.

This fact is already recognized in some countries. The laws in this respect are especially severe in Canada where an owner is not allowed to cut trees around his house without a permission from local authorities asked in a special application with minutely explained reasons.

Countries that lack these limitations (which in a democratic society are liable to public judgment via people's representatives) risk to become subject to catastrophic devastation. This is the case of wood cutting in the Amazon rainforests where fine tree species are being ravaged by home and foreign moneymakers with the connivance of venal officials. Yet the rainforests make up a great portion of oxygen-supplying biomass and depleting them means suffocating all people. Of course, in these cases the society, with its national and international institutions, is obliged to constrain the "private initiative". The boycott of the rainforest timber set up by protection-oriented environmental organizations appears insufficient to save it.

At this point the problem is how far broad public may interfere with private business. People who are against any interference ignore history and cannot foresee the future. Long ago the very concept of property meant that any owner was allowed to use and abuse his belonging, and this was literally the definition in the Roman civil law. However, that definition conflicts with the modern conditions: somebody who wants to build a chemical processing plant on his land abuses his property by polluting the environment, and no right for this abuse is by any means acceptable. Local and global environmental problems can be resolved only by the society, i.e., by interference of the state. Those who sees "socialism" in any interference of state are unconscious of already living under socialism. Any two-alternative approaches — whether to permit or to prohibit any interference — just lead to futile wrangles. Solutions to specific problems always require definite answers.

Pieces of art — painting, sculpture, architectural monuments, etc. — have much in common with environmental wealth. They also can remain in use of many generations of people and can have private owners. However, hardly anybody would think that the owner of art pieces can do with them whatever he wants. In 1687, Turks who conquered Greece set up a powder magazine in the Parthenon and Venetians who wanted to win the "right of property" from the Turks cannonaded and exploded the magazine. Neither Turks nor Venetians had any right for this church, though Greeks, the rightful heir owners of the Parthenon, were not quite aware why should they protect it.

There are apparently things which an international law should protect from barbarians.

Of course, pieces of art have been owned and traded since long ago. It is impossible to steel and hide a church but many paintings stolen by nazi during the second world war survived due to their trade value. Thus even such an apparently strange procedure as monetary evaluation of pieces of art can be a useful protective measure.

Chapter 15

Democracy in the Light of Electoral Procedures

Representative government

The political system called democracy is based on the assumption that legislation and state administration are authorized by the majority of citizens. In the early historic experience, only adult males of a tribe were considered "citizens", and they met at general assemblies to solve crucial issues. However, as far as we know, those assemblies were a mere formality from the very beginning of the historic tradition: The decisions were prepared by the nobility ahead of time, and the assembly, having no right to discuss the proposals, could only approve or disapprove them; the latter happened but rarely. That was the case of the Spartans, the most backward of the Greek tribes where people expressed their opinion by thundering their shields, or the ancient Germans who "voted" by circling around their elders and responding to their suggestions by the buzz of approval or disapproval. In fact all issues were solved privately amongst the nobles, and the wanted attitude of the popular assembly was prepared in advance by manipulations.

We do not know if there had ever been more "democratic" government in the earlier history, but all states of the ancient world initially had popular assemblies. Therefore, the states were small, otherwise it would be impossible to gather all citizens at one place. That is why Aristotle concluded that the dimensions of a state should be such as to make the voice of a town crier audible to all citizens. The idea was an anachronism already in those times since Alexander the Great, Aristotle's pupil, had founded a vast empire ruled in a quite different way. Aristotle might have ignored such an unpleasant fact in favor of the "polis", Greek idea of a city-state.

Popular assembly acquired a much greater importance in Athens of the 5th century B.C., the acme time of the state. It counted then about twenty thousand voting citizens; people made speeches and the decisions were taken by voting during the assemblies. That was the beginning of democracy or "rule by the people". In fact, the political rights of the citizens depended, with rare exceptions, on their estate or property status since the oldest time. Even in Athens full equality of rights held no longer than several decades, and neither women nor the Greek residents of Athens from other states had civil rights, not to speak of slaves. Only twenty thousand of the total 300,000 population of Athens were "citizens". That was the government in the most democratic ancient state.

The democracy of Modern Times arose in England. English barons who forced John Lackland to sign the Great Charter in 1215 never had an idea that anybody besides them might even reason about state affairs. There was no notion of "English people", and the state governance was supposed to be the business of free people: the very word *baro* in old German meant *freeman*; today the German title still sounds as Freiherr (German for *free lord*).

The strive for equality which, according to A. de Tocqueville, had been the driving force of European history for 600 years, rose against estate restrictions and resulted in elective franchise granted first to propertied classes and then to all adults. However, even in the UK universal manhood suffrage was granted as late as in 1884 and women's suffrage had to wait forty years more. Yet, nowadays the "universal, equal and secret suffrage" is regarded as something so much self-evident and indisputable that the very criticism of this institution appears indecent. Therefore, we proceed from the assumption of all citizens having equal civil rights.

Elective franchise means that citizens settle their affairs rather by electing their representatives to the state power to issue laws, appoint and control officials, declare wars, etc. than by meeting at general assemblies. That is representative government. The necessity for such government is evident even if all citizens could really be brought into one place. For example, in Athens where the popular assembly counted several thousand people expressing the opinion for everyone who wanted was already a problem, and keeping order was a still greater problem. Therefore, the Athenians arrived at the idea of "The Council of Five Hundred", or the boule, made up of fifty representatives chosen by lot from each of the ten tribes (phylae). Although the council held general meetings, daily affairs and draft laws were the responsibility of prytanes, smaller groups of fifty men from a tribe, with monthly renewal. That order allowed efficient discussions. People's Court also formed by lot just as jurors are chosen nowadays. Besides, there was something like a government, the office of Strategos, responsible for military decisions and a number of administrative affairs. The strategi were elected by voting rather than by lot as they had to be experts.

The first parliamentary government of Modern Times appeared in England and was later imitated by all countries that claimed to be democratic. The original idea of parliamentary government implied that some special interests were to be represented and not the whole population, and even less that all citizens were represented equally. The earliest parliament pursued only the interests of nobility and clergy, and later also the interests of rich merchants. Nowadays one witnesses a survival of times past, the House of Lords. Moreover, the parliament deputies are still supposed to represent their electoral districts while the "upper chambers" represent the states or national minorities.

However, the most important feature of representative government is associated with the system of political parties which serve as a link between the voters and the government and make nominations to the parliament. Therefore, only the candidates who come to win the favor of party leaders have a chance to be elected. That is why parliamentarians are usually, in a sense, "outstanding" representatives of their voters: sometimes they are more educated, or often rather slier in business than others, and usually stand for the interests of certain groups. In this sense parliaments even now represent more private interests than "a people" as a whole. If the goal of democracy were really a "faithful model" of a society, a lottery would be certainly the best electoral procedure. Of course, an assembly elected in this way would be incompetent and incapable consisting of utterly inexperienced people who are to establish relations and form coalitions, which is indispensable in all human affairs. The party system promotes candidates who have already gained such experience, be it successful or not. In this respect it provides an "improved" representation because novices can hardly maintain effective governance. At the same time, this system spoils representation as businessmen from party circles are most often corrupted.

Our discussion of electoral procedures that follows basically assumes the existence of a party system since, be it good or not, it is inevitable in parliamentary government anyway.

Thus, we assume that for a parliament seat a voter elects a representative from only one of the contesting parties. Suppose for simplicity that one deputy is elected from each district. Since specific personalities are actually designated by parties rather than by voters, the latter can be considered to vote just for a single party. This obviously limits the voters' control over the government. In fact, it often happens that a voter approves some thesis in the program of party *A* (say, a reform) as well as some other thesis in the program of party *B* (say, some ban). Then by voting for the candidate of party *A* the voter supports the former thesis being aware that his[a] candidate will definitely reject the latter thesis; or, vice versa, voting for the candidate of party *B* he supports the latter thesis giving up the former one. In a general case, the election results cannot correctly represent the position of such a voter just because he may support only one party.

It has always been admitted (at least since the Constitution of the United States appeared) that a parliament should also represent people's opinion though it usually promotes the interests of certain groups. This idea — borrowed from the political language of Founding Fathers — is certainly shared by all today's politicians who know the score in political practice and realize that a government should not be "too unpopular". The view finds still more support with the adherents of utopian democracy who wish the government to represent exclusively the people's opinions. In any case, the latter should be taken into account to some extent. Yet, as noted above, people's opinions are biased by the very system of political parties and very often by the requirement to vote only for one party. At this point we put off the question for a while and

[a] For simplicity we use hereafter the personal pronoun *he* for *a voter* be it male or female.

consider how the electoral system – or, more exactly, the way of counting votes – can tamper people's opinion. This problem formulation evidently implies the presence of political parties in the true sense of the word meant as those having a certain program, tradition, and enjoying support from certain groups of population. Not every group of officials or businessmen which can afford engagement in political propaganda is worthy of being called a party. We assume, however, that parties already exist this being the necessary condition for democratic procedures to work.

Proportional and majority electoral systems

The simplest and apparently the fairest electoral system implies that the number of parliament seats allocated to each party is proportional to the number of votes it receives. Yet, historically the earliest (and stable) English electoral system works in a quite different way. The English unwritten constitution assumes that each member of the House of Commons represents the interests of a certain electoral group making up an "electoral district". Since each candidate contests for a parliament seat from a single district he has chosen, the system excludes voting by party lists when people have to vote for a party as a whole represented by the list of its candidates rather than for a specific person. The candidate should be known and popular in his district; originally only the residents of the respective district or those favored by local people of consequence had a chance to be elected. The two-party system has been working in the UK since the eighteenth century, and the electoral race is shared by two main parties of the conservatives (the "Tories") and the liberals (the "Whigs", or the Labour Party of nowadays). With rare exceptions, only the candidates of the two parties compete in each district to represent, as everybody thinks, almost the whole range of political believes of English people. Usually both parties receive a significant percentage of votes in each district, but the votes of minorities are "missed" and do not have any impact on the government. Theoretically, the parties A and B may share votes as 51% to 49% in each district and the party A thus get all seats in the parliament. There were cases when a party received less votes but more seats in the parliament and governed the state till the following elections. In practice a very small advantage of a few percent over the country can give a party overwhelming majority in the parliament. Then, it stays for four years in full control of the government since the latter is responsible only to the parliament. This is the majority system, established also in the United States which borrowed much of the English way of governing.

The majority system is evidently not a very good way to represent the public opinion in a given time. However, an opposition having enough support with people can win at the following elections if the ruling party looses several percent of votes due to mistakes it has committed or just because people got bored with it. A part of population becomes satisfied and the other not too much upset whichever be the winning party unless the two follow quite opposite ideologies, which is usually not the case in the UK or in the US. In general, with such a system, strong electoral groups can influence "their" party and therefore the national politics. A sudden change in public opinion or bad blunders of the ruling party can cause dissolution of the parliament and launch snap elections. This extra-parliamentary mechanism is initiated by a backstage group of experienced politicians but is officially declared by the king or the president.

Efficiency is the basic advantage of the majority system. A parliament with one party in majority supports one-party government and, therefore, consistent policies can be pursued till the following elections: The cabinet does not change or changes only slightly, and the legislation of the ruling party does not impede the political course.

The majority system has been developing for centuries and, as it is often the case of traditional institutions, becomes well adapted to the country it originated in. The proportional system arose only in the nineteenth century in France where radical reforms in the political system through several revolutions razed to the ground the tradition and the new republican order was conceived "out of head" from abstract principles created by "philosophers". Jean-Jacques Rousseau, one of those philosophers, was especially responsible for the course of the French revolution and its legislation.

Inasmuch as the proportional electoral system in France was unfavorable for the two-party system to set in, the French Parliament has commonly comprised many parties, neither of them having the absolute majority. The governments could rely only on fragile coalitions which most often held no longer than several months or, in rare cases, a couple of years. In those conditions of the so-called "cabinet reshuffle", France was unable to conduct a consistent policy between the two world wars and, what is most important, could not set up its military defense and thus doomed itself to defeat in 1940. The proportional system persisted in post-war France as well, with the same consequence of a kaleidoscopic government change. It was the threat of civil war as a result of the "colonial" war in Algeria that made the French consent to the majority system of a mild English type implemented by general de Gaulle. Since then France has had stable government. Italy went through the same evolution after the fall of fascism, but its voting reforms are still underway.

An electoral system is always a compromise between two basic principles of the "People's Rule", or the people's right to influence the government, and the "controllability", or the effectiveness of the government. In fact, politics in general is the art of compromises, as we show below. They are, for example, compromises between irreconcilable voting procedures, though appearing equally "fair", or between the proportional and majority voting systems which both represent essential people's concerns.

Extreme case of a two-party system

It was noticed long ago that the conditions of electoral race in countries with a two-party system gradually make the parties more and more similar and eventually almost undistinguishable in their practice whichever be their historic origin and original ideals. That is what happened in both the United Kingdom and the United States, the states of the classical parliamentary democracy. Of course, the tendency has not yet arrived at its logical end which would evidently cease any principled politics and substitute it by mere competition between public groups for a greater part of the national budget; the competition has the form of peaceful procedures and thus prevents coercive confrontation. This way of arranging public matters is hardly the best one even in the conditions of a stagnant society. Stagnation is generally a too high price for suspending vital problems.

Nevertheless, the two-party system can be efficient to certain limits and is worth considering. Below we investigate how two parties become so much similar in the course of an electoral race assuming as a first approximation that both parties strive uniquely for power. If we further assume, likewise for the sake of simplicity, that the party leaders never abuse their power, a question naturally arises why do they want it. This appears as puzzling as the reason why people play games without any material reward. Anyway, a two-party political game can be useful in resolving not too serious contests for resources.

A more detailed discussion of the topic follows later on in the chapter.

Conventionality of electoral procedures

Leaving aside such complicated questions as shortcomings of the universal and equal suffrage we assume that the principle of universal equality is "fair" and discuss some ways to achieve this equality. Therefore, by "voter" we hereafter mean an "adult citizen". The situation when large electoral groups (or even electoral majority) cannot influence the results of elections and thus miss their votes contradicts civil equality. The electoral law permitting this would be at odds with the idea of democracy and would need revision. We assume for the sake of simplicity that a single chamber of the parliament or a single official (e.g. the president) are to be elected. Of course, all deputies are likewise supposed to have equal rights.

The decision the voters make is whom to elect for the seats specified by the constitution. The constitution outlines the corresponding procedures including the procedure of passing the electoral law. Certainly, the "fairness" (whichever be the meaning of this word) of the constitution and electoral law can be disputed after they have been adopted, but the prescribed procedures should be followed rigorously, otherwise there is no democracy at all. The society unaware of the necessity to adhere strictly and formally to the established procedures is immature and always prone to despotism. In fact, constitutional procedures define the type of democracy in a country.

The constitution postulates a parliament with a certain number of seats, the post of president, etc. The choice of a voting procedure is the following and crucial step in defining democracy because it determines the nature of political power, its activity, and its prestige.

Suppose that a parliament of N seats is to be formed with the given list of candidates greater than N. Without looking into who makes the list of candidates and in which way, and how much this procedure (it is certainly an essential point in the idea of democracy) is controlled by voters, we assume for simplicity that all voters of a country (not divided into electoral districts) or of a single district take part in the elections. The voter chooses N people from the list of candidates a, b, ... , y, z which undergoes reordering according to the electoral rules after the end of certain procedures specified by the electoral law. The first N candidates of the reordered list win seats in the parliament. Several more or less common electoral procedures are as follows.

Relative majority voting. Each voter is allowed to vote for only one candidate, and the candidate who receives the largest number of votes wins the elections.

It may happen that none candidate receives the absolute majority of the votes cast, which accounts for the name of the voting system. Or several candidates may receive equal number of votes and thus have equal reasons to be elected. In this case elections require additional procedures, say, lot.

Simple majority voting. Each voter is allowed to vote for only one candidate and the winner has to receive more than half of the votes cast. If there are more winners than seats, elections continue with additional procedures; if there are more seats, the remaining candidates pass the second round according to the relative majority rule.

This type of voting is most often used in presidential rather than parliamentary elections as there are few candidates and a single winner.

Rated voting. Each voter rates every candidate with a nonnegative integer according to his preferences. If all scores are zeroes, the ballot is discarded. Otherwise all positive scores are normalized through dividing by the sum of all scores to bring the total of all normalized scores to unity. Further counts proceed with the normalized scores to exclude absolute "scale-related" numbers in favor of the relative preference ratio. Sometimes only certain numbers, e.g. integers from 1 to 5, are allowed. Then the scores received by each candidate are summed up and those who received more scores in total are the winners. Moreover, a certain rule prescribes procedures for the case of equal rating. Rated voting is sometimes practiced in sport.

Below we discuss more sophisticated procedures of decision making which may be of practical value. The adopted rules obviously define the representative government or the type of democracy. The rules can be set in a way that only a minor part of voters can influence the results of the elections, at least in single-mandate districts. This way appears "less democratic", and the question thus arises where is the limit to consider voting rules "democratic".

Nominally, the "repertory" of voting procedures includes also the ***"dictator rule"***, which evidently contradicts our intuitive idea of people's rule. It implies that the order of the candidates in the list and thus the membership of parliament is set by a single voter called *first* while other voters have no duties. The name of the rule descends from research of an American economist Kenneth Arrow [*Arrow*, 1963].

Now we try to formulate some general requirements (called axioms following the mathematical terminology) presumably needed for an electoral system to be said democratic.

1. *Universality axiom*. For any two candidates *a*, *b* the voting rules presume three choices: *a* precedes *b*, or *b* precedes *a*, or *a* and *b* have the same rank.

The meaning of the axiom is that the voting rules are universally applicable and consistently give the same result for any two candidates, though not necessarily having ranged them. In such cases the rule is insufficient (as we saw in the previous examples); but we are currently outlining only the necessary conditions of electoral procedures.

2. *Unanimity axiom.* If every individual prefers candidate a to candidate b, the voting result is that a is preferred to b. In other words, unanimous decision of voters should be respected.

3. *Independence of irrelevant alternatives.* The relative order of any two candidates a and b as a result of the voting procedure depends uniquely on the personal preferences of voters rather than on the rank of any other candidate c. It means that a is preferred to b independently of their potential relations to any other candidates.

The latter axiom would appear to represent quite a mature and conscious democracy with voters making their choices independently of the opinions of other people or groups. In fact, the axiom, together with the previous two, which cause no doubt, drive to a paradoxical result. Namely, Arrow proved the following theorem [*Arrow*, 1963]:

The "dictator rule" is the only voting rule to satisfy axioms 1–3.

Without going into the proof of Arrow's theorem we illustrate the root of the problem by a simple model, though, kindred with Arrow's reasoning. Suppose there are only two voters to establish the preference order among a small number of two or three candidates V, B, and M. We try to avoid dictatorship on the one hand and complexities of political life on the other. Yet, the two conditions are incompatible and one is to choose between non-simple political life and dictatorship. Let each voter present his preference list as a string with the most preferred candidate in the first position and the least acceptable in the last position, with or without comma between the positions; the absence of comma means that two candidates are ranked the same. For example, string 1 [V, B, M] means that voter 1 ranked the candidate V as the first, the candidate B as the second, and the candidate M as the third. Suppose we are looking for a good procedure to transform two given preference lists into a third one to be final and depending, of course, only on the order of the initials. For the case of two candidates, the only reasonable non-dictatorial solution is to transform the pair [V, B] (1) and [V, B] (2) into [V, B], and the pair [V, B] (1) and [B, V] (2) into [VB] which means that the contest between candidates ends in a draw if the voters disagree. In fact, if the disagreement in this simple case were resolved by choosing the opinion of voter 1 he would have always profited and thus become the dictator since his decision would be realized in all cases. Therefore the draw is the only non-dictatorial decision to be taken by a true opponent of dictatorship.

Now we slightly complicate the political life by including a third candidate. Let the opinions be distributed as [*V*, *B*, *M*] (1) and [*M*, *V*, *B*] (2). *V* is evidently more preferable than *B* since *V* is more preferable for both voters. Now we try to meet the non-dictatorship rule along with independence axiom 3. Namely, the preference order is chosen for each pair of candidates as if there were no third candidate at all. Mentally excluding the candidate *V* gives [*B*, *M*] (1) and [*M*, *B*] (2) or, according to the non-dictatorship rule, [*MB*], since the two other candidates have the same rank. Excluding candidate *B* gives [*V*, *M*] (1) and [*M*, *V*] (2) or [*VM*] which means equal ranks in another pair of candidates. Combining the two results leads to equality of all three candidates [*MBV*] because they are equal two by two. However, we already found out before that *V* is evidently more preferable than *B* and they are not in the least equal! This example shows clearly that the strivings for non-dictatorship and for perfect simplicity of political life are obviously incompatible. Thus, axiom 3 means primitiveness rather than independence of voters' behavior!

In fact, decisions are never absolutely independent for there are always groups of people joined by common views. Therefore, personal decisions are more or less dependent on the group the person is associated with by birth, education, or his own choice. Of course, these groups not necessarily look like the today's political parties. A society that would satisfy Arrow's theorem can never exist.

Method of statistical sample

Elections are random events which can be described by the probability theory. We start with the simplest case of electing a president from two candidates *A* and *B*. A voter can vote either for *A* or against *A* (by supporting his opponent *B* or not voting at all which is also interpreted as voting against *A*). Then, what is the probability that a random voter (chosen by lot) votes for *A*? According to the frequency definition of probability, the candidate A has the probability p to be voted for if the ratio of the affirmative votes M to the overall number of votes N approaches p for N large enough and becomes still closer to it when N grows. Of course, the procedure implies that a different voter lots off in each trial. The stable frequency $p = M/N$ in these trials means that there is a probability, equal to p, for the candidate A to be preferred. The chosen group of N people (the series) should be really *random*, otherwise the probability laws are inapplicable. For example, a statistical sample (a group) should include approximately equal number of men and women, corresponding to their proportion in the whole population, and uniform distribution of professions, educational level, etc. Generally speaking, a good statistical sample should model the society as exact as possible. Finding such samples is a special art which statisticians are to master, and the sample choice can be checked against the theory of probability.

Furthermore, the probability theory provides methods to estimate the necessary size of the sample N to bring the frequency M/N close enough to the sought probability p. Suppose that the probability of voting for A is known: For example, the elections are over and p can be defined as the ratio of votes for A to the total number of voters. Now consider how the results of elections can be checked. The number M of the A voters from the sample N can range between 0 and N, and for each M there is a probability that exactly M voters vote for the candidate A; let it be $p(M)$. If we know the probability p of voting for A, the probability $p(M)$ for the sample of N people is

$$p(M) = C_N^M \, p^M \, (1-p)^{N-M} ,$$

where the so-called binomial coefficients are given by

$$C_N^M = \frac{N!}{M!(N-M)!};$$

and $N! = 1 \cdot 2 \cdot 3 \cdot \ldots \cdot N$ is the product of all positive integers less than or equal to N, called N factorial (in a similar way $M! = 1 \cdot 2 \cdot 3 \cdot \ldots \cdot M$, $(N-M)! = 1 \cdot 2 \cdot 3 \cdot \ldots \cdot (N-M)$). For example, if a group counts fifty people ($N = 50$), for $p = 2/3$ the probability $p(30)$ of thirty A voters is

$$C_{50}^{30} \cdot \left(\frac{2}{3}\right)^{30} \left(\frac{1}{3}\right)^{20} = \frac{50!}{30!20!} \left(\frac{2}{3}\right)^{30} \left(\frac{1}{3}\right)^{20}.$$

We give the equation for $p(M)$ without proof which is found in any course of the probability theory. Note that direct computing the binomial coefficients is difficult for large M and N and thus other methods of higher mathematics are commonly used instead. What we need is only the general dependence $p(M)$ which is shown in Fig. 1 for $N = 50$, $p = 2/3$. The points $M = 0, 1, 2, \ldots, 49, 50$ are along the M-axis and the values of the function, which is a very close approximation of $p(M)$, are given along the $p(M)$-axis.

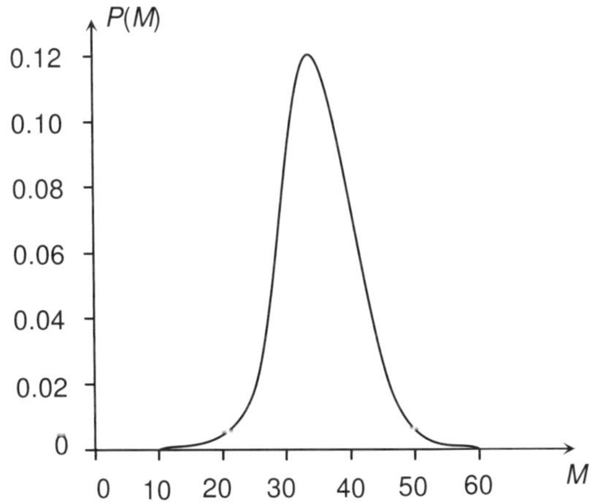

Fig. 1. Gaussian curve of the function $p(M)$ for $N = 50$ and $p = 2/3$.

The bell-shaped curve of Fig. 1 is a Gaussian curve having a great importance in natural sciences and statistics. The maximum $p(M)$ of about 0.12 corresponds to $M = 33$ which is the closest integer to $\frac{2}{3}N$. It means that the most probable result of a poll is that pN members of the selected sample vote for A. At $M \leq 25$ the general probability that no more than a half of the group votes for A is

$$p(0) + p(1) + p(2) + \ldots + p(25)$$

and is vanishing. Exact calculation shows that the sum does not exceed 0.005, and the corresponding probability is below 0.5%.

The probability $p(M)$ can be calculated this way only if the probability p is indeed 2/3, in other words, if 2/3 of all voters did *really* vote for the candidate A. Suppose, however, that a poll among a sample of fifty people shows roughly equal numbers of supporters and opponents of A, say $M = 24$. The probability of this result is below 0.5% at $p = 2/3$. Thus, the poll casts doubt on the declared voting results that A would be put in majority of 2/3.

Of course, falsification is not so easy to discover if the number of "yeas" and "noes" is almost equal. In this case the Gaussian curve peaks near point 25 and the point $M = 24$ falls into the area of rather high values of $p(M)$, which would not contradict the results the election committee reports. Below we show what is to be done in these rather common cases. If, due to falsification, the probable deviation from the true value exceeds significantly the deviation explainable by a small sample size, the results of mathematical control are more trustworthy than official reports even for a sample of only a few dozens.

There is common worldwide practice to check voting results against samples with a size of 2% of the total number of active voters that took part in elections. Samples being large enough in all practical cases, the procedure is well reliable (and its reliability is perfectly amenable to control by mathematical methods). This check is very cheap, and nowadays one can suspect the results of elections where no possibility of independent count is ensured or this count is neglected. The commonly practiced methods of sample control obviously make gross falsification of voting results the deplorable privilege of the most backward countries. More advanced democracies have numerous ways to manipulate the decisions of voters but not their ballots.

The method of statistical sample has other applications beyond the voting control, for instance, predicting election results. This procedure implies pre-election interview of voters selected in a "good" sample ("representative sample" in the terminology of statisticians). The respondents are expected to be sincere though some "timid" people may keep back their unpopular views. Interviewing is an art which requires special experience and has its own rules. According to the probability theory, the results of any type of elections (not only presidential elections where we assumed only two candidates for simplicity) can be predicted to a high accuracy provided that the sample is large enough and properly selected. Sample predictions come from special Institutes for Public Opinion Research and their accuracy increases as the elections approach because the disposition of voters changes with time. Some countries even impose legislation control — especially shortly before voting — on publishing these predictions lest they affect the voters decisions (for example, the followers of a presumably losing candidate can avoid participating in elections). Of course, public opinion polls also prevent falsification of voting results since official information that contradicts consistent sample data published by different institutes can find no credence.[b]

To develop the idea further on, why not cancel elections and use representative sample polls instead? This would make elections much cheaper not in the least changing their results, with the only exception of the case if two candidates share the votes almost equally and voting becomes a lottery with an unpredictable outcome. The method of statistical sample would be in this case nothing but a form of lottery. The use of several independent polls can reduce the risk of corruption among researchers: Falsification immediately shows up if all polls are consistent and contradict the official reports.

Then, it is tempting to make the next step. If a sample is an exact model of the voters tastes and opinions, why a random (skillfully selected!) sample of 500 or 600 people don't serve as a parliament? In fact, if elections indeed pursued the only goal of making an exact model of a society, it would be reached this way with neither big costs nor "ideological campaigns" or mutual defamation of politicians, and moreover, with a solid mathematical background to support the accuracy of "modeling"!

---

[b] There are other prediction methods proceeding from locally specific interests and tastes of voters. For example, reelection (or not) of the sitting President in the United States is predictable from some economy indicators for the last year of his term, i.e., the Americans have long perceived their state authorities responsible for their economic well-being. However, the statistical sample method is universally applicable.

The idea is by no means new. Random sampling is nothing but a good lottery people have practiced since time immemorial. Yet, lot was never used when the elects were expected to have certain skills, experience, or wisdom. Jurors have been chosen by lot till nowadays, since all well-reputed citizens are supposed to be competent in establishing a mere fact of a crime proceeding from available information. We won't linger round the subject and the problem of veto which serves for special interests. In the today's political practice, the right of veto helps suspend urgent problems and supports rather the ambitions than the interests of great powers.

Substituting statistical samples for elections would get stuck on people's mind. For a voter in a democratic society it is important to feel himself a responsible and active participant of state government. That is why personal voting is an essential rite of the democratic culture. Furthermore, no citizen would ever agree to give up the principle of equal responsibility for all public affairs associated with equal suffrage. This ritualized part of public life is hard to change. Many useful improvements to voting procedures, even those not infringing the civil equality, are opposed just for being unusual.

The Zipf–Pareto law

In the course of elections people express their attitude towards politicians or parties by voting for a candidate or for a party. A question arises if there are any regularities to describe the distribution of the votes cast among candidates or parties? If not, votes given for different candidates can be distributed in any proportion or be in any proportion, say, to the turn-out of voters or to the number of invalid ballots. If any regularities do exist, not all distributions of votes are possible.

The results of numerous elections in various states demonstrate a statistical correlation of votes for different candidates and parties described by a simple dependence. If the coordinate axes show, in the logarithmic scale, the number of votes $N(i)$ for each candidate and the rating i of the candidate in the elections, the respective points align with a straight line, to a sufficient approximation:

$$\ln N(i) = A - B \cdot \ln i. \qquad (1)$$

The equation was checked [*Sobyanin and Soukhovolsky*, 1995] through special studies of elections in many countries, from the presidential elections in France (1848) won by Louis Napoléon Bonaparte to the elections of People's Deputies and presidential elections in the 1990s in Russia.

The mathematical result is inherently nontrivial. Experts in physics, chemistry, engineering, demography, ecology, and many other fields that deal with extensive arrays of statistical data are well aware the equation is very general and describes the case of "free competition" for a limited quantity of some conventional "wealth". It was found out that the dependence remains generally the same for any possible diversity of objects, situations, and cause-effect relationships. Provided that a competition is free, its results always align with a logarithmic line in which only the constant A and the slope B can change. Vice versa, if the conditions of free competition are not satisfied, the points inevitably depart from the straight line as far as freedom is violated. This dependence describes, for example, competition of cities for their population in civilized countries. On the other hand, administrative residence limitations (e.g., in Moscow and Leningrad in Soviet time) produce significant deviation from the line of free competition. Free competition leads exactly to the same dependence between the sizes of the biggest fortunes and their ranks in the list of fortunes (where these lists exist). The same law of mass distribution among carnivores works in zoology (in the absence of manmade impact), etc.

This regularity was first noticed by Vilfredo Pareto, an Italian sociologist and mathematician, in his studies of income distribution; the same conclusions were reached later by George Kingsley Zipf, an Americal linguist who studied statistical occurrences of words. Various versions of equation (1) are called the Zipf–Pareto law. The methods associated with the studies of rank distribution are now widely used in linguistics, scientometrics, ecology. The validity of equation (1) in voting means the existence of free competition between all candidates who are free to declare their political views and platforms.

The Zipf–Pareto law in elections means that each candidate, each party, and each electorate group that follow certain voting rules possess their own political platform which does not overlap with the others. Provided that the contesting candidates cover all possible preferences of voters, the percent of voters who seek a candidate beyond the list is rather small and the distribution of votes satisfies equation (1) to a high accuracy. Otherwise, empty niches that may appear in the distribution complicate the analysis.

The parameters A and B in Eq. (1) are calculated through the methods of regression analysis according to the data on the number of votes cast for different candidates or political groups. The parameter A is the logarithm of the number of votes cast for the leading candidate. The preference factor B associated with the slope of line (1) measures the homogeneity of the voters choices. At $B = 0$ the voters have no preferences and all candidates and parties receive equal number of votes. At large B, on the contrary, the outsiders receive much less votes than the leaders (in practice, however, B is almost never above unity). With the assumptions taken, any deviation from line (1) indicates the absence of free political competition. This may result from some external factors such as intimidation of voters by any political or economic constraints for voting (or non-voting) for a certain candidate, or from direct falsification of voting results by election committees at any level. Figure 2 shows a typical rank distribution of voters by the example of elections in Russia. The dependence between the number of electoral groups and their ranks (i.e. the ranks of their candidates) plotted in logarithmic coordinates along both axes is almost linear (Fig. 2).

Distribution of votes cast for different candidates or parties helps to reveal falsification of election results. In the simplest case of stuffing in

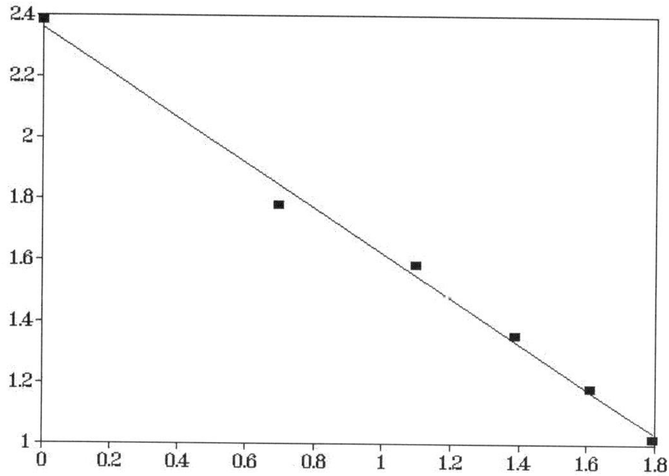

Fig. 2. Distribution of votes at parliamentary elections by party lists, December 12, 1993 (Sverdlovsk region, the Urals, Russia).

favor of some candidate or a party, the rank distribution departs from a straight line, but the points remaining after the pushed candidate or the party are excluded show the theoretically predicted rank distribution. In

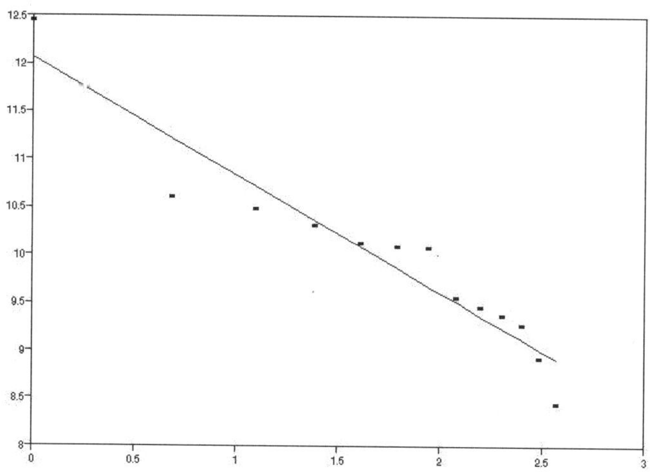

Fig. 3. Distribution of votes at the elections for the seat of the Lipetsk region Governor, spring 1993.

this case the number of stuffed ballots can be measured against the difference between the official data and the theoretical number corresponding to the rank of the candidate obtained from Eq. (1) after the falsified data have been excluded. Figure 3 shows the officially reported distribution of votes cast at the elections for the seat of the Governor in the Lipetsk region (central Russia, not far from Moscow) in spring 1993. The distribution obviously departs from a straight line. Note that the fact of falsification in favor of the winning candidate was established by a court decision in 1995.

The logic of decision making

Election is a specific case of decision making which is the most vital function of all living systems. Human society is such a system making its way through the changing world, and the world survival can depend on the decisions the society makes. Such critical decisions as declaration of war, proclamation of republic, abolition of class privileges, and, finally, the most radical decision of abolition of property rights guide the fates of nations or even whole cultures involving numerous nations.

Many everyday problems the community faces actually require yes/no decisions, called "dichotomous"[c] or binary. Solving binary problems is the common practice in the living world where organisms can be alive or dead, up or sleeping, chase a pray or not, defend or escape. The respective innate decision instructions exist as a binary code which is by no means fortuitous.

Political decisions are obviously not always dichotomous but allow compromises. Compromising is similar to a trade deal if the case is not a matter of principle, such as the ban of drugs which are vaguely classified, or the abortion ban which is not always acceptable. That is why politics is called the art of compromises or, with a blaming attitude, an unsavory business ("sale besogne" in French). The disdain appears

---

[c] The term *dichotomous*, used in many sciences, is of Greek origin (διχοτομία *dichotomia*) which means *division into two*.

reasonable being caused by the fact that political compromises can sacrifice some "sacred" principles.

A dichotomous decision is inevitably associated with discontent of the losing part. The latter often feels itself offended and seeks the guilty or the traitors[d], or more often complains about "unfair play" or bribe, etc. Yet, one part always wins and the other loses even in perfectly fair elections, and the losers have to accept. A worthy behavior of the two parts would be tolerant governing and respect of the rights of minority for the winners and peaceful obedience to the law and preparing to the following elections for the losers.[e]

Which are the practical implications of these ideas? First of all, a voter should be sure the dichotomous election procedure allows him to express his opinion. Or, in the case of a compromise, the way to reach the agreement should be clear and acceptable for every voter whichever be his views.

We can cite the experience of the UK or the US as an example of the former case. With the established two-party system, the "loosing" voters in those countries do not feel much frustration. The balance of forces in a voting district usually means that voters feel themselves supporting a strong and successful party; a party that loses the national elections quite often wins within a district. Then, its supporter can think he did not vote for nothing, because a member of the party *he* supports elected with *his* participation represents local interests of *his* concern which are sometimes put for discussion before elections; this representative is supposed to act in the parliament in a way *the voter* approves. Even though a party loses within the district, its supporters can see an influential and active fraction of the adherents in the parliament; hence, again the "losing voters" feel themselves satisfied. Furthermore, voters can look forward to the following elections knowing that the two parties alternate in power: things can go up if an unlucky voter and his friends are more active. Thus, voters in the two-party system feel their votes

---

[d] That was the case of "patriots" in post-war Germany who could not imagine that their great country (Deutchland über alles) were able to lose the war otherwise.

[e] Note that although the constitution worked in the elections of 1933 in Germany, it was eliminated after the elections, and the obedient German people did not oppose.

have weight and their participation in elections is important. They can even influence to improve the party program and make the party policy more serious.

Many "young democracies" that lack a long parliamentary tradition leading to the two-party system often have numerous more or less small parties which have to cross a certain percent threshold needed for parliamentary representation. Then, the procedure implying that each voter is allowed to vote for a candidate of a single party works against all small parties if the latter are not associated in coalitions. As a result, voters with moderate views who vote for a small centrist party actually miss their votes. Moreover, the few chances to have their representative in the parliament are cut down further by the percent threshold which originally aims at maintaining a stable majority.

Below we consider some new, or at least unusual, voting procedures that help the power to become better representative[f]. Suppose five members are to be elected from the list of more (often much more) than five candidates. In the beginning of the procedure, voters receive ballots to mark five most preferable candidates from the list. Thus the procedure implies the same actions of voters and aims at obtaining the same kind of data as in previous procedures, but the data are used in a quite different way to achieve much better representation of the voters' opinions, though the procedure may seem strange at first glance.

The procedure is performed by a computer count. First, the candidate who received the least number of votes is deleted from the list as the one who is definitely not elected, but then all his supporters are offered a "compensation". Their votes cast for each of the remaining four candidates of the five they chose are counted as 5/4 votes each so that the whole number of available votes remains the same: $4 \cdot 5/4 = 5$. (Of course, this is only a mathematical technique applied to counterweight the disadvantageous position of the minority, and "compensation" is not to be taken literally. Everything happens as if the computer hastened to reward the unlucky voters for their blunder of supporting the outsider!). At the following step, again, the candidate who received the least number

[f] The reader can find the reasons why one of us (V.O.) came to develop the alternative idea of voting procedures at http://www.freewebs.com/nikomo2/index.htm.

of votes (taking into account the compensation) is struck off if there are still more than five people in the list. Then all voters who have four candidates left in their lists of the five they chose are recompensed as above, and the votes of those who have only three left are counted hereafter as 5/3 each to maintain the equal voting potential of all voters. The procedure, easy to perform with a computer and without participation of voters, continues till the final list includes five candidates, which are the winners.

A simple example demonstrates how this unconventional count provides proportional representation whereby a minority receives a number of seats proportional to its size even when the usual statutory procedure would not give it a single seat.

Suppose three people are to be elected in some constituency. Suppose further for simplicity that only two parties contest for the three seats (though the suggested method is advantageous in any case). Let the two parties be "the Left" and "the Right" — whatever be the sense of these names — and assume they have 200,000 and 100,000 supporters, respectively. Let each party have three nominees for three available seats. Then, 200,000 votes are expected to go to each "left" candidate and 100,000 votes to each "right" candidate. As a result of the usual count procedure, the "left" party wins all three seats, but the counting procedure we suggest gives a different result. After the first recount, one "right" candidate (who received the fewest votes) is struck off the list, and the supporters of the "right" party thus have only two candidates left in their ballots. Yet, each of the two receives 150,000 votes, the compensation included. This is still less than the number of votes received by any "left" candidate, and another recount excludes one more "right" candidate. Then the supporters of "the Rights" have a single candidate remaining in their ballots but with 300,000 votes, which is more than any "left" candidate who have only 200,000! Therefore, one "left" goes out in the following recount, and that is the end of the procedure: of three winners of three seats, two are "left" and one is "right", proportionally to the number of their supporters.

The idea of compensation by no means implies that the supporters of outsiders do have the moral right of a reward for their bad luck or inability to predict possible election results. The procedure is a mathematical technique which works as if they had the right and aims at maintaining the proportional representation of voters satisfactory for both the majority and the minority. A strong electoral majority has a majority in a representative body. The case of no such majority requires special consideration in terms of controllability.

The very term *democracy* has different meanings depending on the respective rules of decision making (or votes counting).

Consider now a case when people have no basic dichotomous controversies but just pursue different interests. Political deals in this case can be modeled as something like "fair trade" in which people "buy the right" to solve a problem in their interests. The "market" of problems is assumed to be diverse and the equality of political rights appears as the equality in purchasing capacity, as if a person came to the market of political goods with a certain capital of rights, equal for everybody, and wants to use his capital at his will.

A modeling experiment simulating a "fair market of decisions" was carried out in Krasnoyarsk, Siberia [*Soukhovolsky and Okhonin*, 1995], again using the idea of compensation. Every participant received a standard form with a "set of problems" along with a set of potential solutions for each problem. The participants ranked all suggested decisions by nonnegative integer or fractional numbers. Then the filled forms were processed using a special computer program. At the first step, the scores of each participant were "normalized", i.e., multiplied by a factor to bring to unity the sum of all his scores (all decisions for all problems). Thus the situation resembles a market where a group of buyers possesses equal budgets.

At the next step, the "most unpopular" decision, one with the least sum of normalized scores, was struck off the list. Thereby those who rated that decision to zero had the same scores budget and the others increased, in a standard way, the number of their scores in compensation. This compensation corresponds to the market rules that nobody pays for a mere wish to purchase and undelivered goods are paid back. If several decisions happened to receive the same least number of scores (which is a very rare case), excluded was the uppermost in the list.

Then the procedure was applied again to the reduced list and continued till some problem received a single decision which was supposed to be final. That decision was left in the forms lest a reward were paid to its supporters according to the rule of adding scores for every excluded decision: The luck was supposed to be well earned.

The procedure continued till the decisions were taken for all problems. An additional interviewing program showed that all participants were satisfied with the compromise.

The evident generalization would imply *n* decisions rather than a single decision chosen for each problem. The procedure stops when only *n* decisions remain.

Comment

As we mentioned, the idea of compensation for an unlucky decision, e.g. for a vote loss, is a purely mathematical technique to protect the rights of minorities. It underpinned our previous discussion and appears natural from a rational viewpoint. Indeed, it is impossible to prove that the damage somebody suffers due to his own mistake or disagreement with the majority should be recompensed for someone else's sake. However, the values of our culture are in no way rational. The very idea of universal equality, even though it may seem contradictory to the everyday experience, did not arise for reasons of convenient government or elections but is deeply rooted in human mind and has had insuperable imprint on the course of history for ages. The idea of justice cannot be proved rationally and deduced from optimization of any economic or sociologic parameters.

Even the universal suffrage is nothing but a useful fiction from a purely rational viewpoint. Yet, according to deep humanistic principles every human being has equal rights to participate in public affairs and it would be unfair to deny his wishes, unless they obviously contradict the interests of other people, for example the wish to have a voice and be heard through his representatives in government bodies. This is the origin of the idea of minority rights which are still offended in the electoral practice of many countries. Moreover, every person should have an opportunity to participate in everyday political compromising and have certain "purchasing capacity" in the market of political transactions, for example, as was suggested in the above procedures.

On the other hand, democracy has its pitfalls which neither can be neglected. The victory of German Nazi in the election of 1933 is a tragic lesson for everybody who, following Rousseau, believes in the impeccability of universal suffrage. The ability to use the democratic system of governing is an arduous accomplishment of democracy rather than its starting premise.

A model of competition for votes between two similar parties

Some states considered nowadays as exemplars of democracy historically developed two-party systems with strong competition between two almost equally powerful parties forcing politicians to square up to the opinions of the voters. However, when tried to extend to other countries as is, the corresponding legislation not necessarily works as successfully as in the old traditional democracies. In this respect, they speak about the lack of "political culture and democratic ways". Therefore, a normally running democratic system requires something else besides the written laws. We try to reveal this hidden component by studying competition for votes between two parties. We show below that a competitive two-party democracy that works for the interests of voters is logically possible, but the system fails in certain conditions.

Let all voters be divided into subgroups so that the number of votes each group gives for any party be a function of the party's pledges. In fact, the voters' response to pledges is probabilistic but we assume the groups large enough to allow filtering out random fluctuations to make the response almost deterministic. Assume further the "egoism hypothesis": Let each electorate group respond rather to the pledges that address its expectations than to the party's program as a whole. Finally, we take a "reality hypothesis" meaning that by promising any advantage to some group each party leaves less advantage for other groups, and the latter learn this either from charges of the competing party or by voters' wisdom and education. The voters' preferences are supposed to depend uniquely on the parties' electoral pledges rather than their history, image, and prestige or the charisma of their leaders.

According to the above assumptions, the platforms of both parties are designed to distribute, for example, budget funding among the electoral groups in a way to gain the greatest number of votes. In the same way two companies can compete for clients by investing into advertising that addresses various groups of population. The political case differs as all voters have equal "purchasing capacity" (a vote), whereas companies have to study effective demand and welcome more solvent customers.

Like optimization of marketing strategies worked up by big companies, developing an "optimum program" requires the best knowledge of the functions that describe the voters' response to the electoral pledges. We take some simplifying assumptions about these functions. First, we assume a 100% turn-out rate and thus have to predict only the percentage of voters from any group to support the party *a* as the others vote for *b*. In a general case the function f_1 (percentage of voters in group 1 who support the party *a*) depends on the pledges of both parties *a* and *b*. Hence, we assume for simplicity its dependence

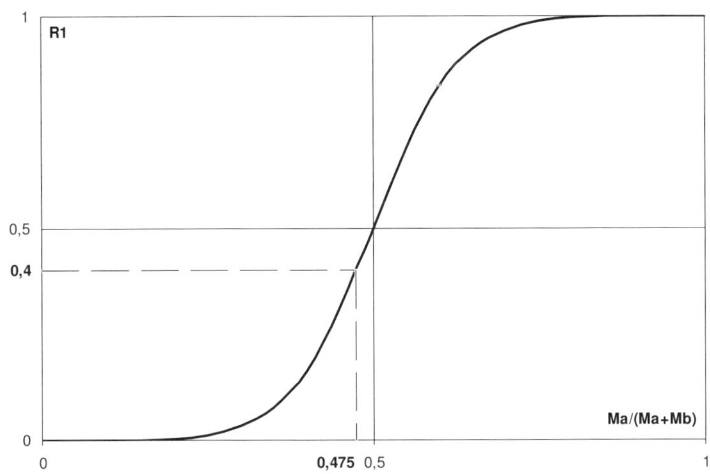

Fig. 4. Model dependence f_1 that shows the percentage of voters in group 1 who support the party *a* as a function of the combined parameter $\mu = Ma/(Ma+Mb)$. *Ma* is the funding for group 1 promised by the party *a* and *Mb* is that promised by the party *b*.

on the combined parameter given by the ratio of funding *Ma* the party *a* promises to group 1 to the total funding promised by both parties *a* (*Ma*) and *b* (*Mb*). Therefore, the functions are assumed to be one-parametric and can be plotted as curves. Figure 4 shows a model dependence f_1 as a function of the combined parameter $\mu = Ma/(Ma+Mb)$.

For some other group 2 the dependence f_2 may look different, e.g., be flatter (Fig. 5).

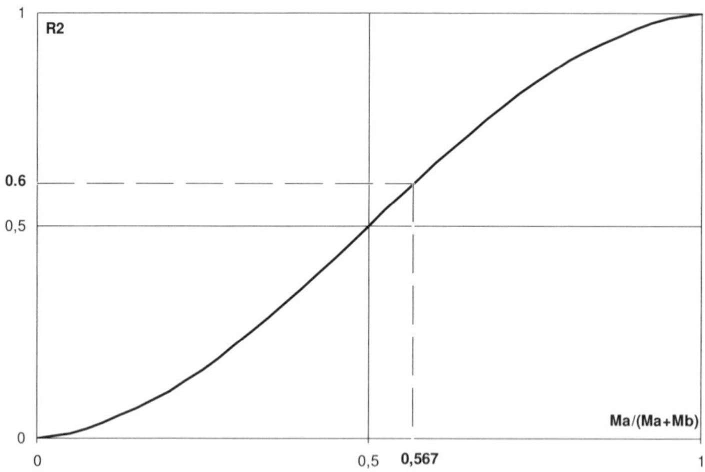

Fig. 5. Dependence f_2 that shows the percentage of voters in group 2 who support the party *a* as a function of the combined parameter μ.

Suppose that at some point both parties promise equal funding to groups 1 and 2, i.e., $\mu = 0.5$. The situation may change in subsequent optimization if at least one party changes its promises in a hope to benefit. This is not always easy and might be impossible. Suppose for simplicity[g] that both groups 1 and 2 are equal in number. Let the party *a* decide to reduce funding promised to group 1 till the level marked by the dashed line in Fig. 4, and μ becomes below 0.5. The released funds are then used to enlarge the promises for group 2 and thus gain new supporters (see the dashed line in Fig. 5). Yet, the party *a* can recruit as many new supporters from group 2 as it loses from group 1 (compare figures 1 and 2). Therefore, the modification of the party program serves for no gain. This situation is obviously possible only if the parameter μ changes consistently in the two groups. In our example, an increase in μ for group 2 is larger than its decrease in group 1 because the relationship for group 2 is flatter. If the funding promised to group 2 increases as much as it decreases for group 1, μ can grow only if the sum $Ma + Mb$ becomes smaller. In other words, the promised funding should diminish proportionally to flattening of the function *f*. With this condition, simple modifications of election platforms give no gain. In fact, this "necessary condition of stability" univocally defines the election platforms. Indeed, the slope of the function *f*, at least in a small vicinity of $\mu = 0.5$, has some constant value (in our case, 4 for f_1 and 1 for f_2). Since the promises should be proportional to this value, groups 1 and 2 are promised, respectively, 4/(1.5+4) = 72.7272...% and 1.5/(1.5+4) = 27.2727...% of the total budget for both groups. The solution is the same for a greater number of groups. If the groups differ in number of voters, funds are distributed per voter, proportionally to the slope of their functions f_i. Thereby the programs of the two parties arrive at an almost perfect coincidence, which is often the case in the political life of some states.

However, if one party steps off the strategy in the above political situation, the other party can immediately profit by and win.

Note that, according to this model of an ideal democracy with two parties having basically the same program balanced to meet the

---

[g] The reader can check that this assumption does not influence the results of our reasoning and thus it was not cited among basic assumptions.

expectations of all electoral groups, the society never breaks up. There is no reason to take the supporters of the opposite party for ideological enemies as the only difference between the parties is that one succeeded more in fitting the balance.

Why then two similar parties should exist? It is because politicians will hardly listen to people unless there is strong competition. The parties should be very similar, they should be two and involved in competition following certain “fair rules” of contest.

The two-party democracy offers a kind of equality rather in terms of intensity of response than in terms of votes. If some group of population remains indifferent to electoral promises (or has zero response intensity), it gets nothing from neither party which in the conditions of competition are forced to offer their promises to more active groups. They are indeed *forced*, as any other strategy leads to failure.

Then, the intensity of public response would be expected to increase because even a small group of voters can receive significant preferences. Such negative effects seem to occur in some modern democracies and may undermine the very institute of competitive democracy when election platforms are unstable and cannot represent in a clear way the actual voters’ concerns.

The cause of instability is that a party can make gains through a great rather than a minor modification of its balanced platform. Minor reduction of funding for some electoral group gives little profit whereas the funds released by a great reduction can serve for effective stimulation of other voters, the deprived group being unable to express its dissent stronger than by casting all votes for the other party.

Imagine, for example, that group 1 makes up 1% of all voters, and the rest 99% is a consolidated group with the intensity (slope) of response x. If the balanced election platforms of the two parties promise, say, 1% of the budget to group 1, one party can suggest to pass the funding to the large group. Thence it looses $0.5 \cdot 1\% = 0.5\%$ of votes it would otherwise receive from group 1: instead of sharing the votes almost equally between the two parties, group 1 unanimously supports the other party which keeps respecting its interests. However, the other, large, group can give about

$$\eta = 99\% \cdot x \cdot \left(\frac{1}{1+0.99} - \frac{0.99}{0.99+0.99} \right) = \frac{0.99\% \cdot x}{2 \cdot 1.99}$$

votes more. The party that modifies its program benefits at $\eta > 0.5\%$ or $x > 2.01$. In the absence of balanced platforms which meet proportionally the interests of all voters it is more profitable to reckon with a large and strongly responding group and ignore the concerns of a small group. The parties are compelled to behave this way to win the race. Therefore, too strong responses of voters are neither socially acceptable.

Generally speaking, acute public response not necessarily brings to the dictate of the majority and ignoring the minority. It can just cause political instability. In fact, large groups are rarely perfectly consolidated but there is rather some hidden disunity to profit by and thus win the election race.

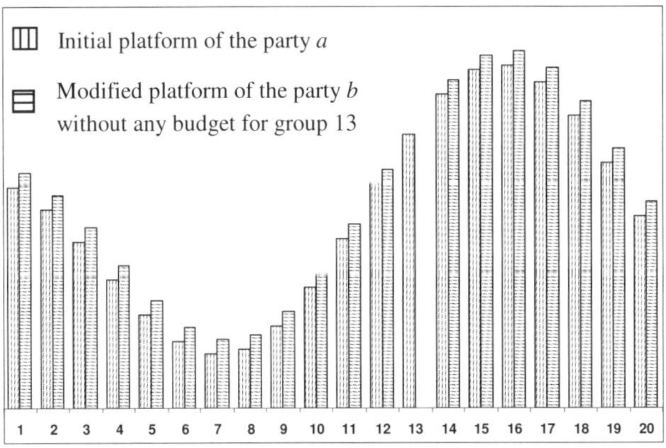

Fig. 6. Initial platform of the party *a* for twenty strongly responding electoral groups.

To better understand the problem, consider a hypothetical society in which all electoral groups respond with utmost intensity, such that any party promising at least a little more than the other immediately receives all votes of the corresponding target group. Suppose the society consists of twenty target groups almost equal in number, and the initial platform of the party *a* looks as in Fig. 6.

Then the party *b* can modify the platform and suggest its own one excluding funding for group 13 and distributing the released sum between the remaining nineteen groups. Thus, each of the nineteen groups receives a little more than before and, according to the assumption of extremely high response intensity, they give all their votes to the party *b* (Fig. 6). Of course, group 13 goes into opposition but is unable to exert any influence since it has only 5% of the votes total. Then, the competing party *a* can suggest another redistribution and exclude, say, group 1 instead of 13 from the budget, following the same simple rule. Thus the nineteen groups, this time including 13 but excluding 1, again profit from the funding the party *b* meant before for group 1. The situation can repeat again and again, and each time the party that makes the last move benefits. Of course, this situation is by no means a good way of governing.

Therefore, democracy is not so simple and even not very reliable government tool. At the same time, the ways to "crack" democratic systems become more and more sophisticated in the today's informational society. On the other hand, the modern science can offer solutions to many problems associated with stability and efficiency of democratic systems.

Not very long ago just few professionals could challenge driving a car on a bad road, but nowadays this is easy for any amateur with the modern systems of automatic control. In the same way, democracy either will advance slowly as far as world nations develop the necessary skills or some appropriate science will provide its breakthrough and allow creating new democracies.

A more detailed discussion of the above ideas requires sophisticated mathematical tools and is expected to be a subject of a special study to be published later.

REFERENCES

Arrow K., 1963. *Social choice and individual values*. NY: John Wiley. Second edition.

Sobyanin A.A. and Soukhovolsky V.G., 1995. Democracy limited by falsifications. Elections in Russian Federation (1989–1993), *Moscow: Human Right Group Publ.*, 266 p. (in Russian).

http://www.freewebs.com/nikomo2/index.htm.

Soukhovolsky V.G. and Okhonin V.A., 1995. *Individual and group choice: methods of description and analysis*. Institute of Biophysics, Krasnoyarsk, preprint 222B.

Conclusion

We wish to end the book with a simple model of human activity. Although strongly simplified, it highlights two extreme types of human behavior: dependent (or service) and independent (or creative). Of course, they are never met just as these but are idealizations of real behavior in which service and creative attitudes make an intricate mixture with various proportions of components. In our model we use the concept of a plain man which follows Adam Smith's idea of a man who "plays fair" but pursues only his self-interest, and this interest is always measurable in terms of money. In fact, it is a "one-dimensional" person being guided by a single numerical parameter. A world inhabited uniquely with these people would be unbearably boring and, as it is easy to predict, would gradually fall into decay. It would not hold long because an individual who has no other motivation besides self-interest sooner or later stops the fair play and tries to evade the rules. Adam Smith, the author of the *Theory of Moral Sentiments*, a treatise on moral philosophy, understood that very well.

The model we use to investigate the human behavior is qualitative. We assume that an individual has the annual revenue q (in money units, say, dollars) and his activity is driven by his interest in the revenue increment Δq (dollars). Then, the dimensionless value $x = \Delta q/q$ can be a measure of incentive, and this is the only incentive for activity of a one-dimensional individual; negative x likewise stimulates the individual's activity: he works more for fear of money loss. Let the corresponding increment in activity be y; its value can be measured in different units, for instance, in units of product the individual manufactures or in units of labor time. The individual is assumed to be engaged in a single type of activity, which is natural for a one-dimensional person, and this activity can be quantified (see Fig. 1 where the x-axis shows the revenue increment x and the y-axis shows the respective increment in activity due to x). Then, the function in Fig. 1 plots the behavior of a "service" man.

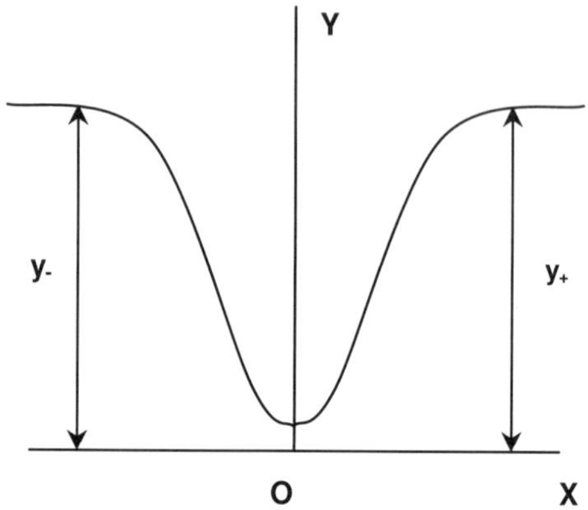

Fig. 1. One-dimensional service behavior.

Activity does not change at $x = 0$, i.e., in the absence of incentive additional to the usual motivation for work. At positive x (incentive), activity grows to some limit y_{max} corresponding to the natural physiological abilities of the individual. At negative incentive, activity likewise rises to reach the same limit but, possibly, following a different law. A one-dimensional man is not supposed to be able of any improvements or innovations which would require other stimuli besides money.

Modeling incentive for an independent and creative person is far more difficult: Creation works spontaneously, without obvious reasons. Yet, one can try to determine the parameters that control creative activity average for people of a certain occupation or a certain hereditary gift. These parameters can include conditions of environment, family status, curiosity, approval or disapproval from one's colleagues and other people around, moral and religious attitudes, and, finally, the presumed community benefit or the opinion of future generations. Being aware of the difficulty or impossibility of quantifying these kinds of incentive, we make an attempt of modeling the creative behavior in a multi-dimensional space with the coordinates u, v, w, ... that define all its motives. The latter can also include money (x) which provide simplest material benefits one is able to buy.

Of course, the "multi-dimensonal" creative behavior is hardly amenable to any straightforward graphic representation. However, it is possible to plot the x dependence taking for constant all parameters u, v, w, ... , which are unmeasurable or only partly measurable in terms of money, e.g., ambitions or the social hierarchy status; their modeler-specified values obviously influence the plot geometry. A heroic person demosntrates the extreme case of independent behavior. This activity can neither be bought for any reward nor suppressed by any punishment. We can cite an example of Jordano Bruno who was burnt in 1600 in Rome, in Piazza Campo dei Fiori. Certainly, one who refused to give up his ideas even under the threat of fire would never do that for money. This case is plotted in Fig. 2 where the activity y is independent of x (though it can depend on other parameters).

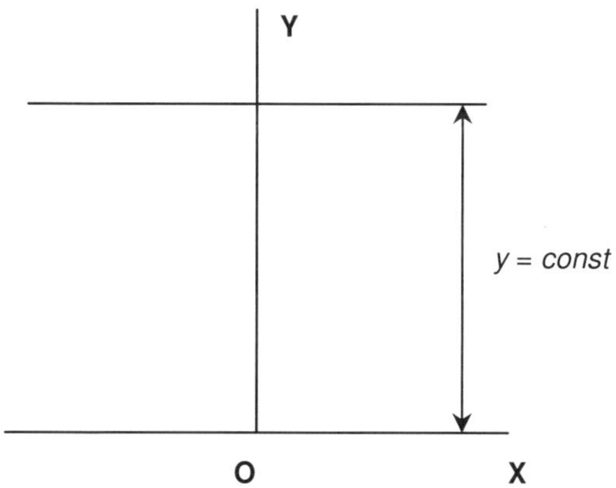

Fig. 2. Multi-dimensonal heroic behavior. y is independent of x

Creative persons are found among many writers, artists, scientists, and among many simple people who value what they want against something different than money. We can cite a number of most striking examples.

The French thinkers of the eighteenth century Enlightenment, leaded by Denis Diderot, accomplished an unprecendented work and published their famous *Encyclopédie*. Albert Einstein, who had worked with noble unselfishness for all his life, advised young scientists to get a job of lighthouse keeper to have enough time for thinking. Louis Pasteur and Ilya Mechnikov risked their lives to create vaccines which saved the mankind from many infectious diseases. Leo Tolstoy did not accept the pay for stories he wrote for people. Alexandre Popov, Russian engineer, the inventor of radio, published his discovery and offered it to the public use. Finally, there are cases when a large group of people is involved in disinterested service for the sake of a high moral ideal — benefit of a nation or the whole humankind. That was, for instance, the Russian intelligentsia.

The attitude of a culture toward this highest creative type of man is a criterion to value the culture itself.

Index